Geometric Dimensioning and Tolerancing

Revised Edition

Gary Gooldy

Prentice Hall
Englewood Cliffs, New Jersey 07632

Library of Congress Cataloging-in-Publication Data

Gooldy, Gary.
    Geometric dimensioning and tolerancing, rev. ed. / Gary Gooldy.
       p.   cm.
    Includes index.
    ISBN 0-13-398959-3
    1. Engineering drawings—Dimensioning.   2. Tolerance (Engineering)
  I. Title.
T357.G66   1995   94–13385
620′.0042—dc20a   CIP

Acquisition Editor: Stephen Helba
Production Editors: Fred Dahl
Copy Editor: Rose Kernan
Designer: Fred Dahl
Project Production Manager: Ilene Sanford / Ed O'Dougherty
Cover Design: Freddy Flake

 © 1995 by Prentice-Hall, Inc.
A Simon & Schuster Company
Englewood Cliffs, New Jersey 07632

Printed in the United States of America

10 9 8 7 6 5 4 3 2

ISBN 0-13-398959-3

Prentice-Hall International (UK) Limited, *London*
Prentice-Hall of Australia Pty. Limited, *Sydney*
Prentice-Hall of Canada Inc., *Toronto*
Prentice-Hall Hispanoamericana, S.A., *Mexico*
Prentice-Hall of India Private Limited, *New Delhi*
Prentice-Hall of Japan, Inc., *Tokyo*
Simon & Schuster Asia Pte. Ltd., *Singapore*
Editora Prentice-Hall do Brasil, Ltda., *Rio de Janeiro*

Contents

About the Author

Gary Gooldy is an active member of the ANSI/ASME Y14.5 sub-committee on Dimensioning and Tolerancing and has been invovled with GDT since 1969. Mr. Gooldy has over 35 years experience in U.S. industry in the fields of design, drafting, and product engineering, and has given countless training seminars in the United States and Europe. His programs are approved by the Society for Manufacturing Engineers and the American Society for Quality Control for recertification credits. Mr. Gooldy is president of GPG Consultants Inc., serving as a trainer and advisor to industry. He has served as an industry advisor to the academic community, and has conducted on-going continuing education programs in cooperation with area universities. Mr. Gooldy was manager of corporate drafting for Cummins Engine Company, Columbus, Indiana prior to retirement in 1987.

The author wishes to express acknowledgement and appreciation to ASME for approval to copy certain figures from the ASME Y14.5 standard. Further, the author wishes to express gratitude to friends and colleagues who have helped in the development and improvement of the Y14.5 standard and, therefore, this work. Of particular note, special thanks to Lowell W. Foster, former Y14.5 chairman and recognized worldwide leader in this subject, for his leadership and friendship over the years.

A special thanks to Janice for sticking by me with support and understanding, and to my little pal, Cody.

Preface

To compete in today's marketplace companies are required to develop and produce products of the highest quality, at lowest cost, and guarantee on-time delivery. A critical element of this process is the communication of the design requirements to the manufacturing/quality areas within the company, as well as to suppliers and customers. ASME Y14.5M-1994 is an engineering communication tool which specifies precise design requirements on engineering drawings which are dictated by the functional relationships of parts and assemblies in their finished state or when ready for the next level assembly. Properly applied, it ensures the most economical and efficient production of the product.

The purpose of this workbook is to serve as supplement material to the national standard ASME Y14.5M-1994 and also as a training aid. This text primarily is for those of basic or limited knowledge of the subject, and is organized and simplified to suit the average user. As this material is advisory, any issues unresolved should be directed to the Y14.5 standard. Readers should have a background in basic math, geometry, blueprint reading, and drafting fundamentals. The figures in this workbook are complete only to the extent they illustrate the point in question, thus any incompleteness may not reflect good drafting/drawing practices.

A further purpose of this workbook is to share with other professionals the author's experiences in application of the principles of the ASME Y14.5M-1994 standard and earlier standards. Engineers, drafters, designers, quality engineers, inspectors, tool designers, gage engineers, manufacturing engineers, or persons from other related fields who must prepare, take action, or interpret engineering drawings, will also benefit from this text.

Geometric dimensioning and tolerancing (GDT) is considered a design specification language, however, it is also a manufacturing and inspection language as well, providing a means for uniform interpretation and understanding between these groups, as well as providing a national/international contract base between customers and suppliers. Examples of this text contain *figures with both metric and inch units* since the system is compatable with either.

At the end of each chapter are exercises which will help summarize the subject covered in each chapter before going on. A final general test appears at the end of the book.

1

Introductory Concepts

To understand where we are and where we are going, we must take a look at where we have been relative to the evolution of standards in this country. Figure 1-1 makes it clear that many issues have been around for quite a while, such as limit gaging (1905), MMC principle (1940), or the military use of symbolism (1945). It is also clear that standardization has been evolutionary, rather than revolutionary. World War II was quite a shock for this country, drove home the need in the military for standards, parts commonality, interchangeability, and reliability, and instantly made the military the world's largest customer with the largest supplier network ever imagined. Even though American industry has not been aggressive in development or adoption of national standard practices, defense contractors have been required to conform to MIL-STDS to acquire and keep contracts, thus forcing their supplier base to conform as well.

Continued development and progress have led increased commonality in worldwide standardization. The United States is a member of the International Organization for Standards (ISO). Other major countries of the world have their own standards, including Australia, Canada, Germany, Great Britain, and Japan. In addition, the countries that form the European Community have developed the EC9000 series standards for their collective market. Other European countries have formed what is called the European Free Trade Association.

1905	The Taylor concept introduced limit gaging for holes and shafts.
1929	Tolerancing first mentioned in French's *Manual of Engineering Drawing*.
1935	ASA Z14.1 *American Standards Drawing and Drafting Room Practice* took 10 years to develop and contained 18 pages.
1940	The *Chevrolet Draftsman's Handbook* introduced MMC in the U.S.
1941	*S.A.E. Aircraft Engine Drafting Room Practices Manual* introduced an elementary form of true position dimensioning.
1945	*U.S. Army Ordinance Manual* introduced symbolism for positioning and form tolerance.
1949	MIL-STD-8 was the first military standard for dimensioning and tolerancing.
1953	MIL-STD-8A authorized use of 7 basic symbols for geometric tolerancing; introduced functional dimensioning methodology.
1957	ASA Y14.5. American Drawing Standard *Dimensioning and Notes*. 35 pages long. No symbolized notes or tolerance expression.
1959	ML-STD-8B became more closely aligned with Y14.5 and S.A.E. and accepted wider symbol usage. Introduction of 3-plane and limits of size concepts. Not approved as an American industry drawing standard.
1963	MIL-STD-8C covered true position tolerancing to a greater degree and introduced the projected tolerance zone concept.
1966	ASA became United States of America Standards Institute and released USASI Y14.5, which introduced tolerancing symbols for cylindricity, runout and profile, changed the definitions of straightness and concentricity. This was also the first unified American standard for dimensioning and tolerancing.
1973	USASI became American National Standards Institute and released ANSI Y14.5, which incorporated all the symbology, established rate tolerances and composite positional tolerancing, introduced LMC, established datum targets, and set forth dual dimensioning methods.
1982	ANSI Y14.5M-1982 ISO Handbook 12 1982 DOD-STD-100

Figure 1-1.   History of geometric tolerancing in America.

Further compounding the issue is the current approval of the North American Free Trade Agreement (NAFTA) by the United States, Canada, and Mexico. The United States and Canada have their own standards, while Mexico uses ISO. While it may be difficult to determine which standards have priority, the agreement, customer, or marketplace will generally determine this.

Let's begin our journey.

Symbol Evolution Summary

The chart in Figure 1-2 shows graphically the development of symbols, beginning with four basic symbols used by the military in the 1940s, and progressing through ANSI Y14.5-1982.

GEOMETRIC AND POSITIONAL TOLERANCE TABLE

CHARACTERISTIC	D-30-1-7 INDEPENDENT	D-30-1-7 DEPENDENT	MIL-STD-8A INDEPENDENT	MIL-STD-8B RFS	MIL-STD-8B MMC	MIL-STD-8C RFS	MIL-STD-8C MMC	USASI Y14.5-1966 RFS	USASI Y14.5-1966 MMC	ANSI Y14.5-1973 RFS	ANSI Y14.5-1973 MMC	ANSI Y14.5M-1982 RFS	ANSI Y14.5M-1982 MMC
CONCENTRICITY (COAXIALITY)	◉ A xxx	A xxx	◉ A xxxTIR	◉ A xxxTIR	◉ A Ⓜ xxxTIRⓂ	◉ A xxxTIR	◉ A Ⓜ xxxTIRⓂ	↗ A xxx ◉ A xxx DIA	⊕ A xxx DIA	↗ A xxx TOTAL ◉ A ⌀xxx	⊕ A Ⓜ ⌀xxxⓂ	⌰ xxx A ◉ ⌀xxx A	⊕ ⌀xxxⓂ A Ⓜ
PERPENDICULARITY (SQUARENESS)	⊥ A xxx	A xxx	⊥ A xxx	⊥ A xxx	⊥ A Ⓜ xxxⓂ	⊥ A xxx	⊥ A Ⓜ xxxⓂ	⊥ A xxx	⊥ A Ⓜ xxxⓂ	⊥ A xxx	⊥ A Ⓜ xxxⓂ	⊥ xxx A	⊥ xxxⓂ A Ⓜ
PARALLELISM	∥ A xxx	A xxx	∥ A xxx	∥ A xxx	∥ A Ⓜ xxx Ⓜ	∥ A xxx	∥ A Ⓜ xxx Ⓜ	∥ A xxx	∥ A Ⓜ xxx Ⓜ	// A xxx	// A Ⓜ xxxⓂ	// xxx A	// xxxⓂ A Ⓜ
SYMMETRY	℄ A xxx	A xxx	—	≡ A xxx	≡ A Ⓜ xxxⓂ	≡ A xxx	≡ A Ⓜ xxxⓂ	≡ A xxx TOTAL ⊕ A Ⓢ xxx TOTAL Ⓢ	⊕ A xxx TOTAL	≡ A xxx ⊕ A Ⓢ xxxⓈ	⊕ A Ⓜ xxxⓂ	⊕ xxxⓈ A Ⓢ	⊕ xxxⓂ A Ⓜ
ANGULARITY	—	—	∠ A xxx	∠ A xxx	∠ A Ⓜ xxxⓂ	∠ A xxx	∠ A Ⓜ xxxⓂ	∠ A xxx	∠ A Ⓜ xxxⓂ	∠ A xxx	∠ A Ⓜ xxxⓂ	∠ xxx A	∠ xxxⓂ A Ⓜ
INTERRELATED CHARACTERISTICS (DIA AND SURFACE AT RIGHT ANGLE TO COMMON AXIS)	◉ ⊥ A D xxx	A D xxx	—	◉ ⊥ A D xxx TIR (TOL RFS) ◉ ⊥ AⓂ D xxx TIR	◉ ⊥ AⓂ D xxx TIRⓂ (DATUM MMC)	◉ ⊥ A D xxx TIR (DATUM RFS, MMC) ◉ ⊥ AⓂ D xxx TIR	◉ ⊥ AⓂ D xxx TIRⓂ (TOL RFS)	↗ A D xxx (DATUM MMC) ↗ A Ⓜ D xxx (TOL RFS)	⊕ A D xxx DIA	↗ A B xxx TOTAL (DATUM MMC) ↗ A Ⓜ B xxx (TOL RFS)	⊕ A Ⓜ B ⌀xxxⓂ	⌰ xxx A B (DATUM MMC) ↗ xxx A Ⓜ B (TOL RFS)	⊕ ⌀xxxⓂ A Ⓜ B
FLATNESS	—	—	∼ xxx	⌢ xxx	—	— xxx	—	▱ xxx	—	▱ xxx	—	▱ xxx ▱ xxx .010/.25X.25	—
STRAIGHTNESS	—	—	⌢ xxx	⌢ xxx	⌢ xxxⓂ	— xxx	— xxxⓂ	— xxx	—	— xxx — ⌀xxx	— ⌀xxxⓂ	— xxx — ⌀xxx — ⌀xxx ⌀xxx/xxx	— ⌀xxxⓂ
POSITION (HOLES)	—	xxx	⊕ A xxxDIA	⊕ A Ⓢ xxx DIAⓈ	⊕ A xxxDIA	⊕ A Ⓢ xxx DIAⓈ	⊕ A xxxDIA	⊕ A Ⓢ xxx DIAⓈ	⊕ A xxx DIA	⊕ A Ⓢ ⌀xxxⓈ	⊕ A Ⓜ ⌀xxxⓂ	⊕ ⌀xxxⓈ A Ⓢ	⊕ ⌀xx[illegible] A Ⓜ
POSITION (SLOTS)	—	—	—	⊕ A Ⓢ xxx TOTALⓈ	⊕ A xxx TOTAL	⊕ A Ⓢ xxx TOTALⓈ	⊕ A xxx TOTAL	⊕ A Ⓢ xxx TOTALⓈ	⊕ A xxx TOTAL	⊕ A Ⓢ xxxⓈ	⊕ A Ⓜ xxxⓂ	⊕ xxxⓈ A Ⓢ	⊕ xxxⓂ A Ⓜ
ROUNDNESS (RADIAL)	—	—	—	—	—	○ xxx ON R	—	○ xxx	—	○ xxx	—	○ xxx	—
ROUNDNESS (DIAMETRAL)	—	—	—	○ xxx NOTE 1	—	○ xxx NOTE 1	—	—	—	—	—	—	—
PROJECTED TOL ZONE	—	—	—	—	—	⊕ xxx DIAⓈ TOL ZONE PROJECTED xxx	⊕ xxx DIA TOL ZONE PROJECTED xxx	⊕ xxx DIAⓈ TOL ZONE PROJECTED xxx	⊕ xxx DIA TOL ZONE PROJECTED xxx	⊕ A ⌀xxxⓈ xxxⓅ	⊕ A ⌀xxxⓂ xxxⓅ	⊕ ⌀xxxⓈ A xxxⓅ	⊕ ⌀xxxⓂ A xxxⓅ
PROFILE OF A LINE	—	—	—	—	—	—	—	⌒ xxx	—	⌒ xxx	—	⌒ xxx	—
PROFILE OF A SURFACE	—	—	—	—	—	—	USED NOTE	⌓ xxx	—	⌓ A xxx	—	⌓ xxx A	—
RUNOUT (CIRCULAR)	—	—	—	—	—	—	—	↗ A B xxx CIRCULAR	—	↗ A xxx	—	↗ xxx A	—
RUNOUT (TOTAL)	—	—	—	—	—	—	—	↗ A B xxx	—	↗ A xxx TOTAL	—	⌰ xxx A	—
CYLINDRICITY	—	—	—	—	—	—	—	⌭ xxx	—	⌭ xxx	—	⌭ xxx	—
POSITION (COMPOSITE FEATURE CONTROL SYMBOL)	—	—	—	—	—	—	—	—	—	⊕ A B C ⌀xxⓈ A ⌀xxⓈ	⊕ A B C ⌀xxⓂ A ⌀xxⓂ	⊕ ⌀xxⓈ A B C ⌀xxⓈ A	⊕ ⌀xxⓂ A B C ⌀xxⓂ A
RUNOUT (TWO DATUM DIAMETERS WITH COMMON AXIS)	—	—	—	—	—	—	—	—	—	↗ A-B xxx TOTAL	—	⌰ xxx A-B	—

Figure 1-2.

Key points include:

- The use of concentricity and TIR simultaneously through 1966.
- Position symbol (1953).
- Evolution of straightness and flatness.
- Changes in the feature control frames and combined applications.
- The development and introduction of runout (1966).
- Addition of profile tolerancing (1966).
- Development of symmetry (1945), dropping symmetry (1982).
- Composite tolerancing (1973).
- Supplemental symbols added (1982).

Symbolism continues to develop and improve, and much progress has been made in gaining agreement with ISO on standardized symbols and their application. (See Figures 1-3, 1-4, 1-5, and 1-6.)

Feature Control Frames

Now that we are familiar with symbols, we may arrange them to form symbol sentences. This is accomplished by the use of Feature Control Frames. The symbols are placed in a specific order, within the feature control boxes: the geometric control first, the tolerance zone shape and tolerance second, and the datum(s) references third. Refer to Figure 1-7 and ASME Y14.5M1994 for further information on feature control frames. These feature control frames are used throughout this text.

Fundamental Rules

We should also be familiar with some basic drawing and quality control rules. These rules have been extracted from the following standards:

- ASME Y14.5M-1994: Dimensioning and tolerancing
- ANSI B4.4: Inspection of workpieces
- ANSI B89.6.2: Temperature and environmental controls
- ASME Y14.2: Line conventions and lettering

These fundamentals (see Figure 1-8) should be understood because, with the general rules in Figures 1-15 through 1-18, they form the foundation for the Y14.5 standard.

SYMBOL FOR:	ASME Y14.5M	ISO
STRAIGHTNESS	—	—
FLATNESS	▱	▱
CIRCULARITY	○	○
CYLINDRICITY	⌭	⌭
PROFILE OF A LINE	⌒	⌒
PROFILE OF A SURFACE	⌓	⌓
ALL AROUND	↙○—	↙○— (proposed)
ANGULARITY	∠	∠
PERPENDICULARITY	⊥	⊥
PARALLELISM	//	//
POSITION	⌖	⌖
CONCENTRICITY (concentricity and coaxiality in ISO)	◎	◎
SYMMETRY	⌯	⌯
CIRCULAR RUNOUT	•↗	↗
TOTAL RUNOUT	•⌰	⌰
AT MAXIMUM MATERIAL CONDITION	Ⓜ	Ⓜ
AT LEAST MATERIAL CONDITION	Ⓛ	Ⓛ
REGARDLESS OF FEATURE SIZE	NONE	NONE
PROJECTED TOLERANCE ZONE	Ⓟ	Ⓟ
TANGENT PLANE	Ⓣ	Ⓣ (proposed)
FREE STATE	Ⓕ	Ⓕ
DIAMETER	Ø	Ø
BASIC DIMENSION (theoretically exact dimension in ISO)	[50]	[50]
REFERENCE DIMENSION (auxiliary dimension in ISO)	(50)	(50)
DATUM FEATURE	•[A]	• or •[A]

• MAY BE FILLED OR NOT FILLED

(ASME Y14.5M-1994)

Figure 1-3.

Definitions

The definitions on pages 6 and 9-12 are similar to those of Y14.5, with some shortened for simplicity. These definitions are also part of the foundation for Y14.5, and are terms used throughout the standard. These definitions should be understood and followed for continuity. ASME Y14.5M-1994 is the final authority for precise wording. (Other brief definitions are contained in the Glossary at the end of the book.)

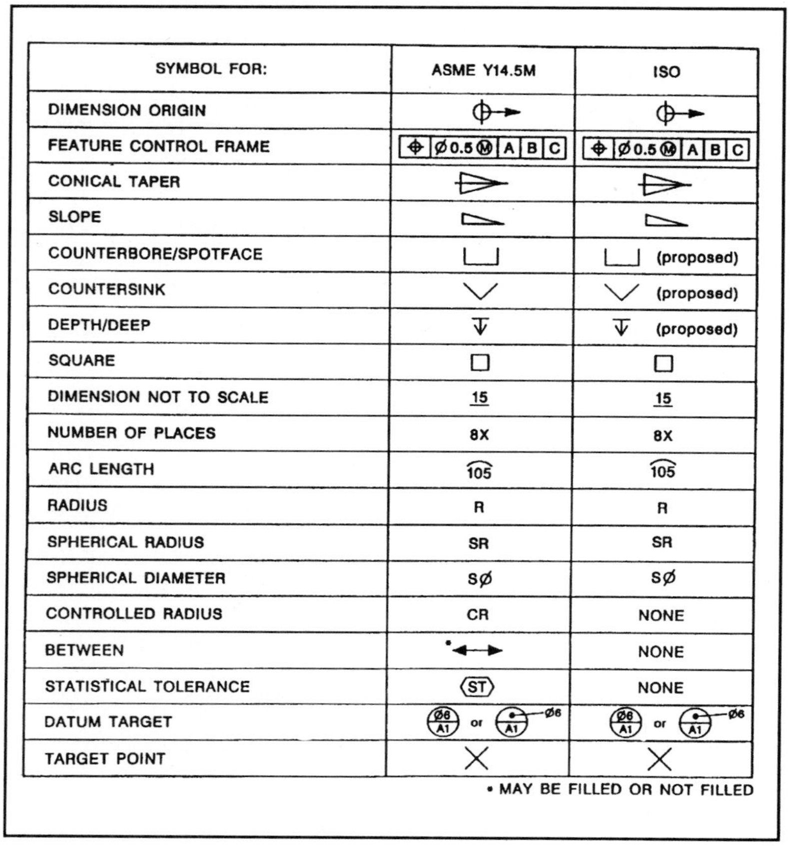

SYMBOL FOR:	ASME Y14.5M	ISO
DIMENSION ORIGIN	$\phi\!\!\rightarrow$	$\phi\!\!\rightarrow$
FEATURE CONTROL FRAME	⊕ ⌀0.5Ⓜ A B C	⊕ ⌀0.5Ⓜ A B C
CONICAL TAPER	▷	▷
SLOPE	◿	◿
COUNTERBORE/SPOTFACE	⌴	⌴ (proposed)
COUNTERSINK	⌵	⌵ (proposed)
DEPTH/DEEP	⤓	⤓ (proposed)
SQUARE	□	□
DIMENSION NOT TO SCALE	<u>15</u>	<u>15</u>
NUMBER OF PLACES	8X	8X
ARC LENGTH	⌒105	⌒105
RADIUS	R	R
SPHERICAL RADIUS	SR	SR
SPHERICAL DIAMETER	S⌀	S⌀
CONTROLLED RADIUS	CR	NONE
BETWEEN	•◄—►	NONE
STATISTICAL TOLERANCE	Ⓢ̲Ⓣ	NONE
DATUM TARGET	⌀8⁄A1 or •⁄A1 ⌀6	⌀8⁄A1 or •⁄A1 ⌀6
TARGET POINT	✕	✕

• MAY BE FILLED OR NOT FILLED

(ASME Y14.5M-1994)

Figure 1-4.

Actual Local Size. The measured value of any opposed points or cross section of a feature.

Actual Mating Size (Envelope). The value of the actual mating envelope, i.e., the smallest possible cylinder about an external feature, or the largest possible cylinder within an internal feature.

Actual Size. The general term for the size of a feature; includes the actual mating size and actual local size.

Basic Dimension. A numerical value that describes the exact theoretical size, shape, location, and other characteristics of a feature, datum, or target. The value is contained within a box.

Centerline. Generic term for axis.

Datum. A theoretically exact point, line, or plane derived from the feature counterpart.

Datum Feature. An actual part feature, including feature irregularities of flatness, circularity, cylindricity, and straightness.

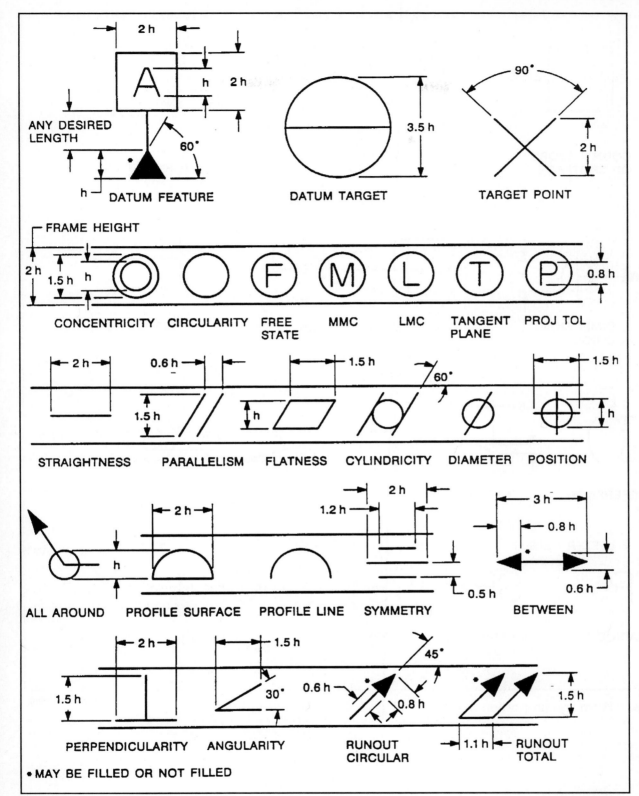

Figure 1-5. Form and proportion of geometric tolerancing symbols.

(ASME Y14.5M-1994)

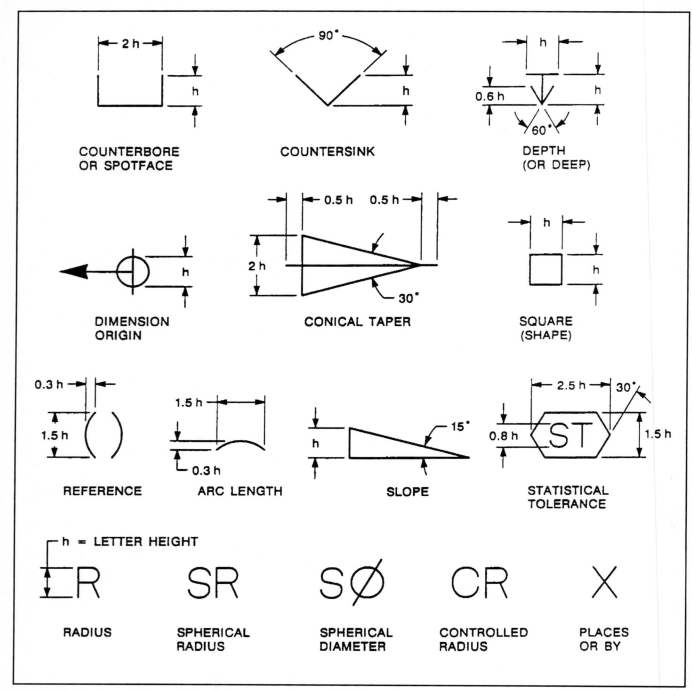

Figure 1-6.   Form and proportion of dimensioning symbols and letters.

(ASME Y14.5M-1994)

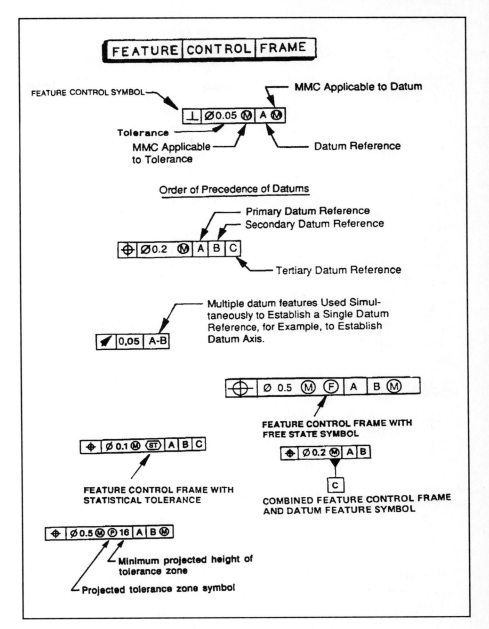

Figure 1-7. Feature Control Frame

Datum Target. A point, line, or area on a part feature used to establish a datum, or otherwise to ensure consistency and repeatability in processing or measurement.

Dimension. A numerical value to define part characteristics: size, geometric, and the like.

1. All dimensions are toleranced (except reference, min., max., or stock).
2. Dimensioning shall be complete.
3. There should be no more dimensions than necessary.
4. Dimensions are functional with one interpretation.
5. Drawings define parts without manufacturing methods unless those methods are engineering requirements.
6. Nonmandatory processing dimensions and information are noted [Mfg. Data].
7. Dimensions are shown in profile views to visible outline.
8. Unless otherwise specified, lines shown crossing at 90° are 90°.
9. Dimensions apply at 20° C (68° F) unless otherwise specified.
10. Dimensions apply in the free state unless noted (except non-rigid features).
11. Dimensions and tolerance application must be clear on plated or coated parts.
12. Decimal dimensions shall be used (except for commercial or standard stock applications).
13. There should be no double "dimensions."
14. Dimensions *not* to scale are underlined.
15. Dimensions should appear outside object lines.
16. The length, width, or height of the tolerance zone is equal to the length, width, or height of the feature.
17. Dimensions and tolerances apply at the drawing level specified and are not mandatory at any higher level (assembly drawing) altered by process or assembly.
18. The drawing must specify the standard of reference (ASME Y14.5M 1994).

Figure 1-8.   Fundamental rules of dimensioning.

Feature. General term to describe a physical portion of a part, i.e., hole, surface, slot, and so on.

Feature Axis. A straight line that coincides with the centerline of the feature counterpart.

Feature Centerplane. A true plane that coincides with the centerplane of the feature counterpart.

Feature of Size. A feature with opposed feature elements, i.e., cylindrical, hexagonal, spherical, or two parallel plane surfaces.

Free State. The condition of a part absent of any restraining forces.

Full Indicator Movement. The total movement of the indicator device for measuring surface error.

Functional Gage: A term to describe a dedicated (limit) gage that will receive the part with no force applied. Elements of the gage may be variable.

Least Material Condition (LMC). The condition in which the feature contains the least material (i.e., the part weighs the least).

Maximum Material Condition (MMC). The condition in which the feature contains the maximum material (i.e., the part weighs the most).

Median Line. An imperfect line that passes through all the center points of all cross sections of a feature, normal to the actual feature mating size envelope.

Median Plane. An imperfect plane that passes through the center points of all cross sections of a feature, normal to the actual feature mating size envelope.

Nominal Size. Term for general identification.

Reference Dimension. A dimension without tolerance for information only.

Regardless of Feature Size (RFS). The condition to indicate that a datum reference or geometric tolerance applies at any size increment within the size limits.

Resultant Condition. The boundary generated by the collective effects of size, material condition, geometric tolerance, and any bonus due to the feature departure from its specified material condition.

Simulated Datum. A datum simulated by processing or inspection equipment surfaces: surface plates, tool centers, datum simulators.

Tangent Plane. A theoretical exact plane that contacts a feature surface at the high points on the surface.

Tolerance, Bilateral. A tolerance that exists in two directions from a specified dimension.

Tolerance Bonus. An increase in form, orientation, or position tolerance allowed, equal to the feature's departure from the stated material condition size.

Tolerance, Geometric. The general term for the categories of form, orientation, profile, location, and runout tolerances.

Tolerance, Statistical. The assignment of tolerances to component parts (features) equated to formulas relative to the square root of the sum of the tolerances.

Tolerance, Unilateral. A tolerance that exists in only one direction from a specified dimension.

True Geometric Counterpart. A theoretically perfect boundary (virtual condition mating envelope) of a feature.

True Position. The theoretically exact location of a feature established by basic dimensions.

Virtual Condition. A constant (worst case) boundary generated by the collective effects of a size feature's specified material condition and the geometric tolerance for that material condition.

Tolerance

Until the early 1920s, tolerance was generally considered a result of machine capability, along with operator skill. Tolerancing did not generally appear on engineering drawings, and the use of fractions, plus single or multiple surface finish marks (*ff*), confirms this: the more finish marks, the closer the control. We realize dimensions must have limits. Limits determine the fit with mating parts. Mating fits are defined and classified in ANSI B4.1 and B4.2.

Tolerance impacts both quality and cost, from the initial design concept through the manufacturing and service development process, to the customer. Tolerance impacts new products initially in new tooling costs, process time, equipment maintenance costs, quality planning, and audit costs, process control costs, scrap/salvage costs, and measurement capability or gaging costs. Hidden costs due to tolerancing inefficiencies or misunderstanding may be found in virtually every product if we look hard enough.

To avoid these hidden costs, we should be able to defend and justify design tolerances both logically and mathematically:

- Tolerance should be based on function and on functional relationships.
- Tolerance values have been standardized in classes and fits, as noted earlier, as well as in handbooks and industry standards for such things as fasteners and O-rings. Society manuals and handbooks also contain many standardized tolerances.
- Tolerances may come from previous designs, using similar parts as reference, which may or may not repeat past errors.
- As a last resort, we may pull a tolerance out of the air, such as ±.010, or use title block drawing tolerances.

Quality at least cost is the goal, along with a logical and well defined design that is mathematically sound.

Design Considerations and Tolerance Distribution

Print tolerances are values used by various agencies in the design-through-manufacturing process. Tolerances also represent engineering requirements that must be met for product acceptance. For any given tolerance, gage designers (or other inspection process) have 5% of the engineering drawing tolerance for the gage, with another 5% for gage wear (see ANSI B4.4 Inspection of Workpieces). Tooling manufacturers generally have 15-20% of print tolerance for tooling, leaving manufacturing approximately 70% of the print tolerance. This 70% is based on new equipment; as tools and fixtures wear, adjustments must be made. We rely on the process control system and line gaging, or audits, to maintain conformance to these tolerances. Figure 1-9 graphically illustrates a typical breakdown of tolerances to achieve 70% tolerance with 99.73% repeatability ($\pm 3\sigma$). Quality standards often require the center of the curve of distribution be within $\pm 10\%$. Many industries are working toward $\pm 4\sigma$ (99.95%) and closer to achieve quality goals.

The key point is that manufacturing does not have available all the tolerance specified on the print. The distribution of tolerance may differ with a given business or industry, but the principles are similar.

Tolerances and allowances for many size ranges have been standardized. ANSI B4.1 (inch) or B4.2 (metric) contains the classes of fits and tolerances. There are five classes of fits:

1. RC-running and sliding fits
2. LC-location clearance fits
3. LT-location transition fits
4. LN-location interference fits
5. FN-force fits

Each of these classes is subdivided into groups. These groups include:

- *9 RC clearance classes of fits.* The amount of tolerance is proportional to the size of the bore/shaft.
- *11 LC location classes of fits.* These range from small clearance (LC1) to liberal clearance (LC11). LC fits provide assembly clearance but are not intended for moving/mating surfaces.
- *6 LT classes of fits that may result in either a clearance or interference condition.*
- *3 LN classes of fits.* For location of parts when a clearance is not desired.
- *5 force or shrink classes of fits.* These provide constant pressure (press) through a range of sizes.

Tables for the varied classes and tolerances may be used for mating part features, but, regardless of method of determining tolerance, the tolerances are to be anchored by function.

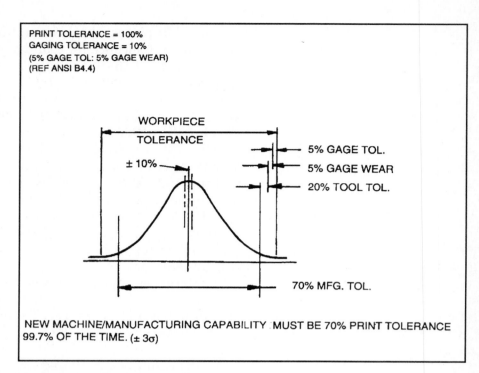

Figure 1-9.   Design considerations.

Gaging Tolerance—5% workpiece tolerance (size)—5% wear allowance.
Measurement Allowance—within limits of given gaging tolerances location & size
Random Error—varied, unpredictable error ... not possible to account for ... only can fix limits within which error probability will occur.
Form Control Error—allowed 50% gaging tolerance.
Workpiece Limits Interpretation (Size)
    Holes—largest perfect imaginary cyl.
    Shafts—smallest perfect imaginary cyl.
Choice of Method—Based on Purpose, Volume, Shape, Size, Geometry, Location, Direct Meas. vs. Limit Gages.
General Principles—No part of the surface shall infringe on the MMC envelope ... separate form controls must be correlated with size dimensions.
Inspection by Measurement Tolerance—must be held within the workpiece size limits.
Ref. Stds.—ANSI B4.2 Pref Metric Limits & Fits
            ANSI B47.1 Gage Blanks
            ANSI B89.1.9 Precision Gage Blocks
            ISO 463 Dial Gauges Reading
            ISO 1302 Technical Drawings-Surface Texture
            ISO 3650 Gauge Blocks
            ASME Y14.5.1M Mathematical Definitions for Y14.5M

Figure 1-10.   Excerpts from ANSI B4.4M–Inspection of workpieces.

Tolerance/Fit Study

Let's explore tolerancing further, by looking at a typical fit of a shaft and bore from the tables in ANSI B4.1 (inch). Let's choose an RC6 (medium running fit) and apply the table tolerances, observing the results.

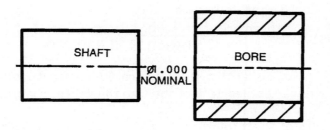

NOMINAL SIZE	RC6		
	TOLERANCE 0.000		
	CLEAR.	BORE H9	SHAFT e8
1.000	1.6 4.8	+2.0 0	-1.6 -2.8

ANSI B4.1

Expressed as Class Fit	Bore	1.000H9
	Shaft	1.000e8

Unilateral Tolerance   Bore $1.000 \begin{array}{l} +.002 \\ -.000 \end{array}$

                      Shaft $1.000 \begin{array}{l} -.0016 \\ -.0028 \end{array}$

Limits   Bore   1.000
                        1.002
          Shaft   .9984
                        .9972

In the product development/manufacturing process of a product, all areas involved share the preceding tolerances. For example, gaging and inspection has 10% of the available tolerance for gaging/measurement allowance per ANSI B4.4. Tooling and fixturing tolerances account for approximately 15 to 20%, leaving manufacturing 70% of the available tolerance, as previously discussed.

Tolerance Distribution

Shaft tolerance allowance	.0012	10%	gage	.00012
		20%	tool	.00024
		70%	manufacturing	.00084

Bore tolerance allowance	.002	10%	gage	.0002
		20%	tool	.0004
		70%	manufacturing	.0014

Now apply geometric tolerances of form, orientation, profile, location, or runout. As a general guideline, geometric controls should not exceed one-half the size tolerance of a feature.
Example:

Total bore tolerance = .0012

      therefore | — | .0006 | (max.)
                  SURFACE ELEMENT

Total shaft tolerance = .002

      therefore | — | .001 | (max.)   or | ↗ | .001 | A |
                  SURFACE ELEMENT

Also, generally, the surface finish (texture) of a feature should never exceed 10% of the size tolerance. Example:

Total shaft tolerance = .0012 × 10% = .000120 μ in. Ra, or $\sqrt[120]{}$

(Generally this value would be rounded down to a more preferred size such as $\sqrt[64]{}$.)

There is more discussion of surface texture at the end of the book on pages 231-232 with more complete information found in ANSI B46.1 and ANSI Y14.36.

This has been a simplified study of a typical dimensioning and tolerancing process, but not all projects will be as simple. The basic elements of the process—preferred fits, tolerance distribution, functional analysis, determination of geometric and surface texture controls—will still exist. We therefore have a starting point for establishing ground rules.

Let's apply these principles to some typical examples. First is an assembly of two parts, as shown in Figure 1-11. Using unilateral tolerancing helps to illustrate the fits and tolerance conditions that occur at MMC. From the calculations shown, we see that the minimum clearance (both parts at MMC) of the mating datum features is 0.5. Also, the minimum clearance of the related features is 0.5 at MMC. Therefore, there is a total combined orientation or location tolerance available of 1.0. This tolerance may be distributed among the features of both parts in any combination that does not exceed a total of 1.0.

For this example, we distribute the tolerance equally between parts, 0.5 to each, as shown.

If it were possible to make perfect parts, the fit of the two parts might look like Figure 1.11b. The positional tolerance would be evenly distributed, and the axes would be aligned. We know, however, it is not possible to make totally perfect parts, and therefore a positional error can exist, as illustrated in Figure 1-11c. The amount of clearance at the feature(s) surface (gap) at MMC is equal

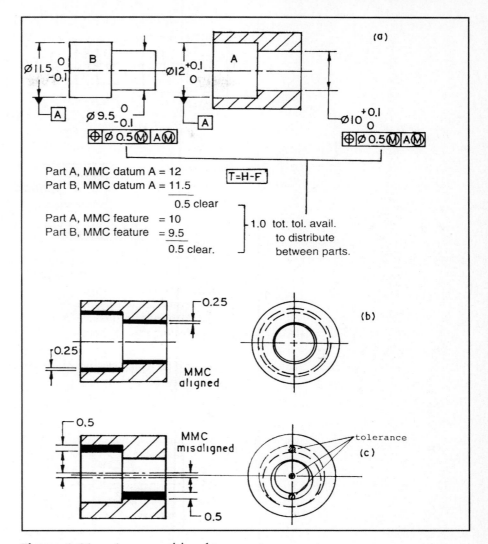

Figure 1-11.   An assembly of two parts.

to the total position tolerance of those features. Figure 1-11c represents both parts at MMC that have used all available position tolerance, thus illustrating a worst-case MMC assembled fit.

Further tolerance is allowed if one or more features and/or
datum features depart from the MMC size limits. Figure 1-12a
illustrates a possible combination of LMC sizes and resulting
bonus tolerances. This combination results in the loosest fit possible for the two parts. Figure 1-12b further illustrates a combination of tolerances, including orientation (parallelism) error,
which could occur at LMC, yet still allow assembly of the two
parts because of bonus tolerance and datum shift.

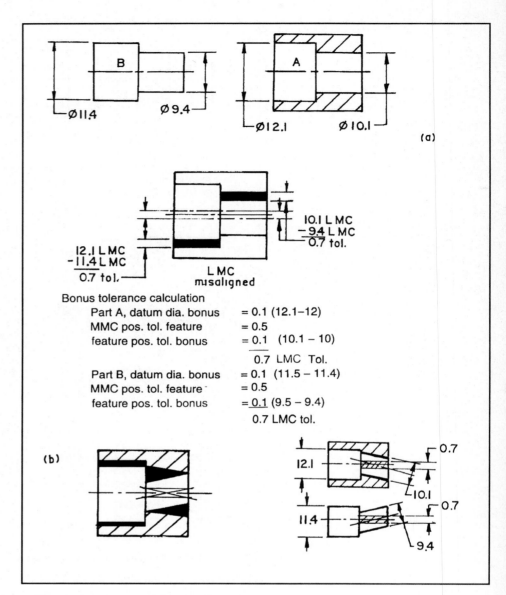

Figure 1-12.   LMC fit conditions.

The previous study and principles may also be applied to assemblies of three aligned diameters, as shown by Figure 1-13. For multidiameter assemblies, with *all diameters free to float*, the tolerance and resulting worst-case fit are found in the same fashion as with two diameter assemblies. The total tolerance is distributed among all the features. If the two feature diameters are offset to the maximum allowed location tolerance at MMC, a line-to-line fit would result, allowing no additional tolerance from the datum feature. See Figure 1-13b.

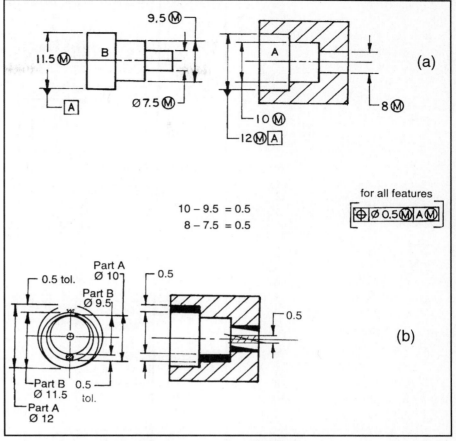

Figure 1-13. Assembly of three aligned diameters.

In Figure 1-14, the datum is a threaded feature which "locks" the assembly in place. The available tolerance must be halved and distributed between all diameters of both parts. Although we will do more on this in the position tolerance section, this exercise serves to introduce us to fundamental tolerancing principles. Remember: A datum feature that is a feature of size, such as a shaft or bore, may also have a location or orientation tolerance/consideration, relative to the total assembly. If the datum features involved create a fixed fit, such as a press fit, threaded fit, or line fit, they must be using some of the tolerance that will not be available, as datum "float," to other features.

Fixed Assembly Formula   Tolerance = $\dfrac{\text{Bore} - \text{Shaft}}{2}$

$$T = \frac{H - F}{2}$$

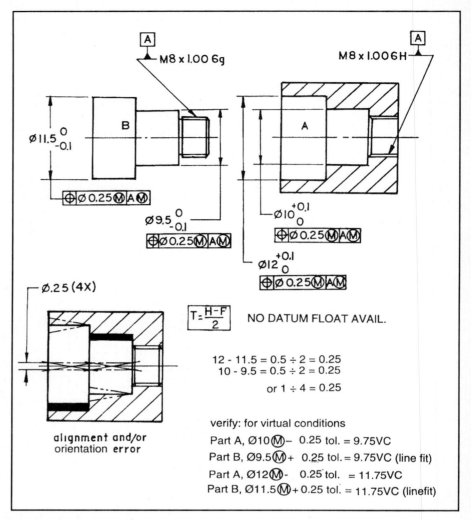

Figure 1-14.   Threaded feature datum.

Free Assembly Formula   Tolerance = Bore – Shaft

$$T = H - F$$

where T = tolerance, H = hole/bore, and F = fastener/shaft.

Tolerance Expression

In the Y14.5 standard, we are told that:

• When using limit dimensioning, the high limit (maximum value) is placed above the low limit (minimum value).

- When tolerancing in a straight line (.740-.760), the low limit is expressed first, regardless of the feature type (internal vs. external).

This practice appears awkward, and it can be hard to remember. Further, since MMC conditions give the closest fit and are the basis for functional gage designs, it would make sense to show the MMC limit as the top dimensional size limit. In addition, many machinists work to the MMC limits first so as to have the remaining tolerance as needed for tool wear or rework. For these reasons, the author has used the "MMC-limit-first" approach on most examples in this workbook using limit dimensions.

Putting the MMC limit above, or first, is not in accordance with ASME Y14.5. However, it does serve to highlight the MMC conditions of material features, and points us toward "preferred limits and fits" of ANSI B4.1 and B4.2.

Note: *This approach is used as a training and learning aid only, and it is not recommended for any other purpose.*

The conventions pertaining to the number of decimal places carried in the tolerance shown in the following paragraphs shall be observed.

Millimeter Tolerances. Where millimeter dimensions are used on the drawings, the following applies.

Where **unilateral tolerancing** is used and either the plus or minus value is nil, a single zero is shown without a plus or minus sign.

Example:

$$50^{\;0}_{-0.02} \quad \text{or} \quad 50^{+0.02}_{\;0}$$

Where **bilateral tolerancing** is used, both the plus and minus values have the same number of decimal places, using zeros where necessary.

Example:

$$50^{+0.25}_{-0.10} \quad \text{not} \quad 50^{+0.25}_{-0.1}$$

Where **limit dimensioning** is used and either the maximum or minimum value has digits following a decimal point, the other value has zeros added for uniformity.

Example:

$$\begin{matrix} 40.45 \\ 40.00 \end{matrix} \quad \text{not} \quad \begin{matrix} 40.45 \\ 40 \end{matrix}$$

Inch Tolerances. Where inch dimensions are used on the drawing, both limit dimensions or the plus and minus tolerance and its dimension shall be expressed with the same number of decimal places.

Examples:

.500	not	.5
.498		.498
.500±.005	not	.50±.005
$500^{+.005}_{-.000}$	not	$.500^{+.005}_{0}$
30.0° ±.2°	not	30°±.2°

Basic   *Basic*

| 50 | | .500 |

with   with

⊕ | ∅0.15 Ⓜ | A | B | C   ⊕ | ∅0.05 Ⓜ | A | B | C

Dimensions are implied as absolute (10.5 implies 10.5000000).

General Rules

Three general rules are important to the further principles in Y14.5, and should be reviewed carefully. Note Rule #1 does not apply to commercial stock or nonrigid parts as defined in Y14.5, section 6, and as shown in Figure 9-10 in Section 9. Note also that Rule #1 does not *always* have to apply, and may be avoided or disclaimed, as shown in Figure 1-16. Rule #1 does not control the relationships of features, as shown in Figure 1-17. Note: ISO does not apply the effects of Rule #1, but accomplishes this result by the use of the symbol Ⓔ "envelope."

Rule #2 (Figure 1-18) has been revised from previous standards:

- USASI Y14.5-1966 implied MMC for tolerances of position.
- ANSI Y14.5-1973 allowed an optional application.
- ANSI Y14.5M-1982 required the use of a modifier Ⓜ, Ⓛ, or Ⓢ, for positional tolerances.

In the ISO system, all geometric controls are, and have always been, implied RFS. With this revision to Rule #2, the ASME and ISO standards are in agreement, relative to this principle.

The rule for screw threads, gears, and splines remains unchanged.

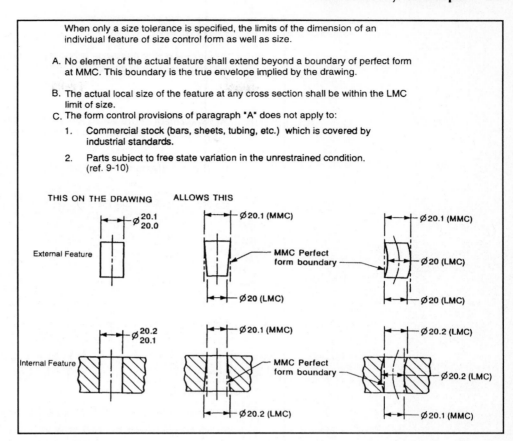

When only a size tolerance is specified, the limits of the dimension of an individual feature of size control form as well as size.

A. No element of the actual feature shall extend beyond a boundary of perfect form at MMC. This boundary is the true envelope implied by the drawing.

B. The actual local size of the feature at any cross section shall be within the LMC limit of size.

C. The form control provisions of paragraph "A" does not apply to:

　　1.　Commercial stock (bars, sheets, tubing, etc.) which is covered by industrial standards.

　　2.　Parts subject to free state variation in the unrestrained condition. (ref. 9-10)

THIS ON THE DRAWING　　　ALLOWS THIS

External Feature　　ø 20.1 / 20.0

ø 20.1 (MMC)　　MMC Perfect form boundary　　ø 20 (LMC)

ø 20.1 (MMC)　　ø 20 (LMC)　　ø 20 (LMC)

Internal Feature　　ø 20.2 / 20.1

ø 20.1 (MMC)　　MMC Perfect form boundary　　ø 20.2 (LMC)

ø 20.2 (LMC)　　ø 20.2 (LMC)　　ø 20.1 (MMC)

Figure 1-15.　　Rule #1.

Material Condition

Even though MMC was used in the *Chevrolet Handbook* of the 1940s, consideration of feature material condition (size) has generally not been applied in determining tolerances for part acceptance. We have been accustomed to applying tolerances at their face value (i.e., ±.010), without regard to size (RFS). There are three material conditions to consider: MMC, LMC, and RFS. MMC and LMC have symbols, while RFS does not, as it is implied unless otherwise specified (Rule #2). MMC and LMC, when applied to feature controls, will allow a bonus tolerance as the feature departs from its MMC or LMC size limit. This amount of departure may be added to the specified geometric tolerance as a bonus tolerance (more on this later). RFS is the condition that applies to coordinately dimensioned drawings, as the coordinate system has no provisions for applying MMC, LMC, or bonus tolerance. (See Figure 1-19.)

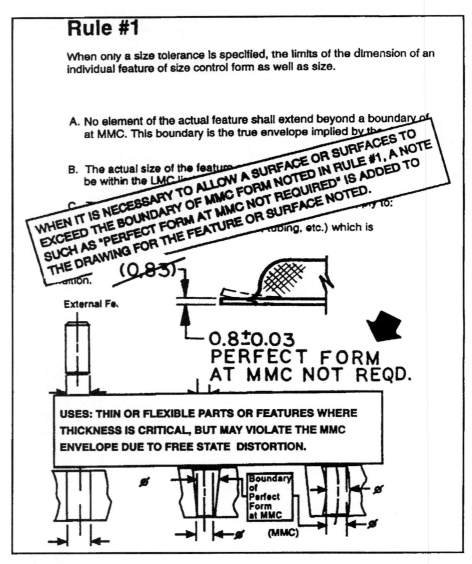

Figure 1-16. Rule #1.

Summary

To sum up the general section:

- We have reviewed history and standards evolvement.
- Symbols have world-wide meaning and application.
- We can link symbols in feature control frame sentences.
- There are fundamental rules, definitions, and general rules that are the foundation of Y14.5.

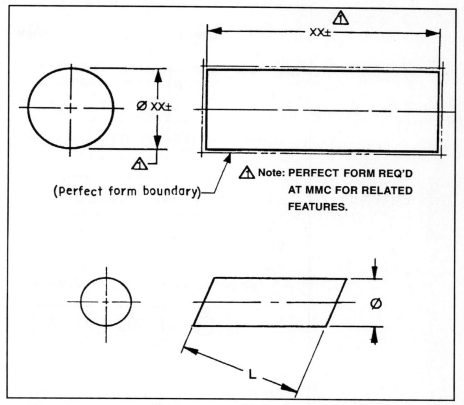

Figure 1-17. Rule #1 applies to a single feature, not to the relationship of features. This relationship may be controlled by a note, as shown, or by the use of datums.

- Tolerances should be logical, justified, and defendable.
- Material conditions can impact tolerances.

In developing a good design, it is well to consider size controls first (Rule #1), then form controls, since Rule #1 controls form to a degree, as well as size. Orientation, profile, runout, and position controls are considered after we have dealt with size and form. (See Figures 1-20 and 1-21.)

Complete Exercise 1-1, and then explore form tolerances in Chapter 2.

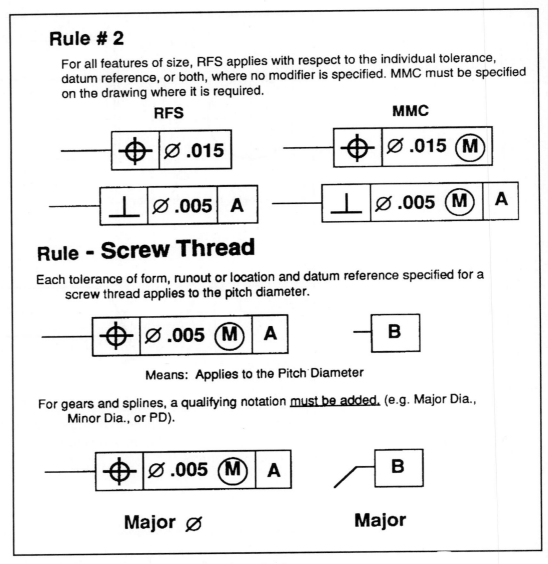

Rule # 2

For all features of size, RFS applies with respect to the individual tolerance, datum reference, or both, where no modifier is specified. MMC must be specified on the drawing where it is required.

Rule - Screw Thread

Each tolerance of form, runout or location and datum reference specified for a screw thread applies to the pitch diameter.

Means: Applies to the Pitch Diameter

For gears and splines, a qualifying notation **must be added.** (e.g. Major Dia., Minor Dia., or PD).

Figure 1-18.   Rule #2 and screw thread rule.

Maximum Material Condition

Symbol: Ⓜ

Abbreviation: MMC

The condition where a feature of size contains the maximum amount of material within the stated limits of size, for example the high limit size of a shaft and the low limit size of a hole.

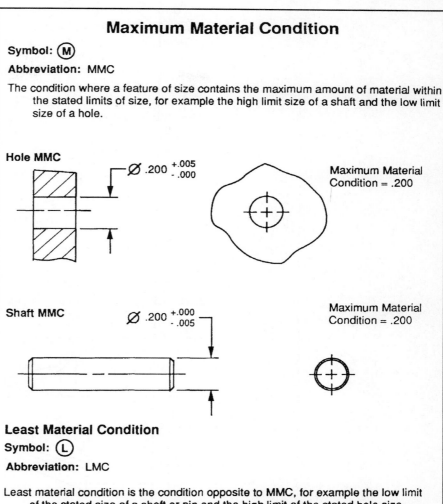

Hole MMC

$\emptyset$ .200 $^{+.005}_{-.000}$

Maximum Material Condition = .200

Shaft MMC

$\emptyset$ .200 $^{+.000}_{-.005}$

Maximum Material Condition = .200

Least Material Condition

Symbol: Ⓛ

Abbreviation: LMC

Least material condition is the condition opposite to MMC, for example the low limit of the stated size of a shaft or pin and the high limit of the stated hole size.

Regardless of Feature Size

Symbol:

Abbreviation: RFS

The geometric tolerance applies at any increment of size of the feature within its size tolerance.

Figure 1-19. Material conditions.

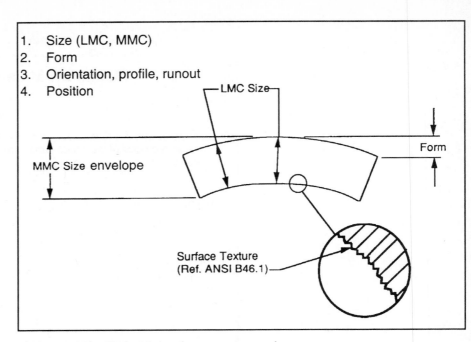

1. Size (LMC, MMC)
2. Form
3. Orientation, profile, runout
4. Position

Figure 1-20. Selection of proper control.

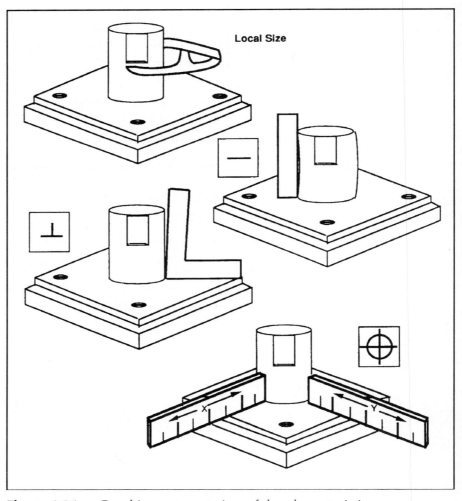

Figure 1-21. Graphic representation of the characteristics.

EXERCISE 1-1. GENERAL DIMENSIONING

1. General Rule #1 controls:
   a. form as well as size of feature.   T   F
   b. the relationships of shaft ends to the diameter.   T   F
2. General Rule __ implies RFS is applied to runout controls.
3. Prior to ANSI Y14.5–1973, position controls implied MMC.   T   F
4. ISO drawings imply MMC for position controls.   T   F
5. All dimensions on a drawing must have a tolerance.   T   F
6. Drawing callouts referring to screw threads apply to the pitch diameter.   T   F
7. Identify the figures in the feature controls frame below:

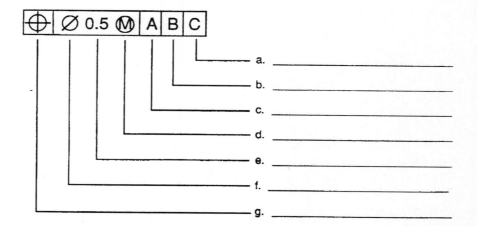

a. _____

b. _____

c. _____

d. _____

e. _____

f. _____

g. _____

8. Identify each of the geometric characteristic symbols below.

2

Form Tolerances

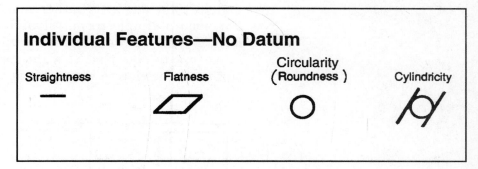

Individual Features—No Datum

Straightness   Flatness   Circularity (Roundness)   Cylindricity

Figure 2-1.   Tolerances of form.

Now that we've covered the fundamentals, general rules, definition of terms, and we have a general foundation of basic principles in place, let's begin with the most basic of geometric controls.

The most basic element is a *point* in space. Two points will create a *line*, while a line and third point (or three points) will create a *plane*. The intersection of two planes also will form a line (*centerline* or *axis*). Points equidistant from an axis create a *circle* and two circles on the same line create a *cylinder*.

The descriptions in Figure 2-1 comprise the most elementary controls in the Y14.5 standard. They are classified as tolerances of form:

- straightness (line);
- flatness (plane);
- circularity or roundness (circle); and
- cylindricity (cylinders).

The common thread between all form tolerances is that they are all individual features with no relationship to any other feature. They are easily described, but more difficult to measure and evaluate as there are no datum references. Form tolerances generally are considered a refinement of size. When applied to a surface or surface element (line), they must be contained within the size limits per Rule #1 unless otherwise stated, "perfect form at MMC not required."

Straightness

Straightness is a condition where an *element of a surface or an axis* is a straight line. Straightness tolerance specifies a tolerance zone within which an axis (derived median line) or the considered element must lie. The control of all line elements is *straightness*. The control of surface line elements is two dimensional, namely *height* and *length*. There are many designs where the effect of matched line elements rather than surfaces is the desired intent. For such a simple term, "straightness" can be a very complex form control, as we will find in the following illustrations. It is important to note that straightness tolerance applies only to the surface in the view in which it is specified, as shown by Figure 2-2.

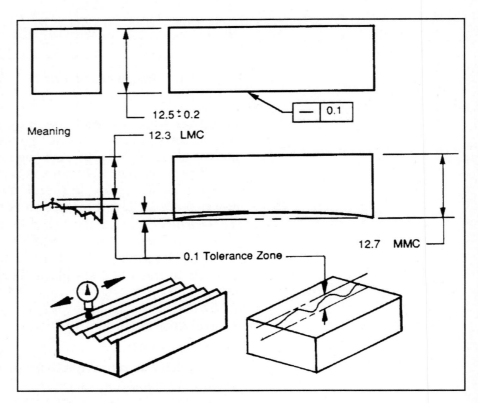

Figure 2-2.   Straightness of surface elements.

Straightness-Cylindrical Features

In Figures 2-3a and 2-3b, we see a different placement of the feature control frame. In Figure 2-3, the frame is separated from the diameter dimension with a leader line pointing directly to the surface. In Figure 2-3b, the frame is below and directly linked with the diameter/size dimension. This frame placement creates two totally different designs. The tolerance zone for the surface line elements of Figure 2-3a is at the surface, as shown, while the tolerance zone for Figure 2-3b is at the feature axis. Because the zone in Figure 2-3a is at the surface, the tolerance must fall within the limits of size per Rule #1. Because the tolerance zone of

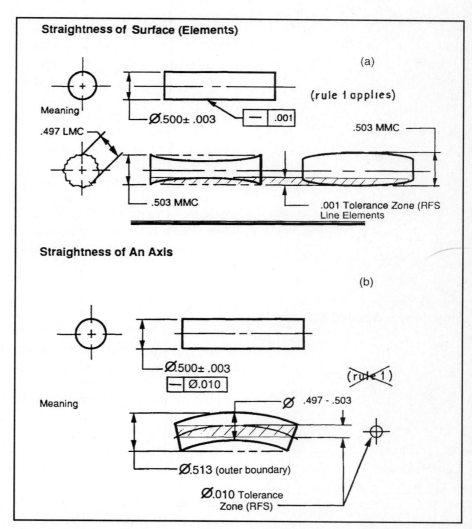

Figure 2-3. Straightness of cylindrical surface elements and an axis (both RFS).

Figure 2-3b is at the axis, the axis is allowed to be bent within the .010 diameter zone, and the combination of maximum size plus axial straightness will allow the feature to violate the boundary of perfect form at MMC (Rule #1). This set of conditions is termed *outer boundary*. We will discuss this later on in the text.

With cylindrical parts, we must first understand that the placement of the feature control frame can change the design intent dramatically. Straightness is applied RFS, unless otherwise noted. Where a straightness tolerance control is used in conjunction with an orientation or position tolerance, the specified straightness tolerance *shall not* be greater than the specified orientation tolerance values. When not used in conjunction with an orientation or position tolerance, and when the straightness is applied to a feature axis, the straightness tolerance may be greater than the size tolerance. In this instance, the MMC limits of size may be violated. (See Figure 2-6.)

As shown in Figure 2-4a, straightness may also be applied MMC. In this example, the axis must be straight within a diameter tolerance zone of .010 when the feature is at the MMC size of .503. The figure illustrates that as the size departs from MMC, we gain a bonus tolerance value we can add to the original straightness tolerance. The virtual condition or mating size (collective effects of MMC size, and geometric tolerance) remains the same (a .513 diameter). The hole in the mating part would have to be at least .513 diameter to insure clearance fit. Look again at Figure 2-3b. The straightness control is applied RFS and the tolerance is constant at .010 diameter, so the outer boundary, or mating size, is variable from a .513 to .507 diameter. With RFS, tolerance is constant while the mating size is variable. With MMC, the mating size is constant while the tolerance zone is variable.

Straightness Applied Per Unit Length

If our design cannot tolerate an abrupt step, as shown in Figure 2-4b, we may use tolerance per unit length as well as the total tolerance. The straightness symbol is entered once and is applicable to both total and per unit straightness.

Centerplanes

In the past, straightness controls have been applied to *centerplanes* as shown in Figure 2-5b. As the definition of straightness deals with surface line elements or axes, this application of straightness control to a centerplane may be argued as incorrect.

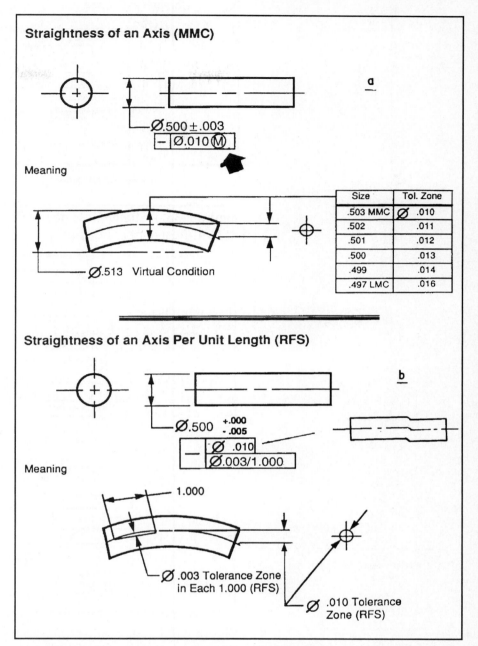

Figure 2-4. Straightness of an axis (MMC) and an axis per unit length (RFS).

If the design intent is to control the centerplane of a feature, such as a keyslot, the use of true position may be considered. However, if we need to control surface elements, straightness controls applied to the feature surface along with true position of the centerplane may be the proper control. The design should dictate the controls specified.

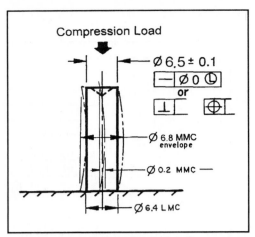

Figure 2-5a. Straightness of an axis at LMC (resultant condition).

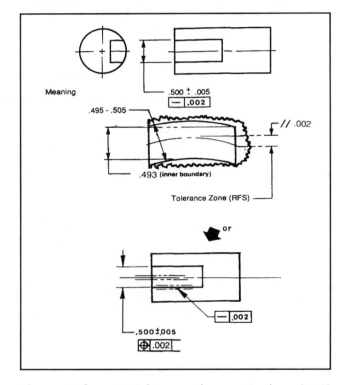

Figure 2-5b. Straightness of a center plane (RFS).

Virtual Condition

Virtual condition is the boundary generated by the collective effects of MMC or LMC size limits and any other geometric tolerance. Keeping in mind the definition of virtual condition, we can recognize the effects of applying a geometric control, such as straightness, perpendicularity, parallelism, or position to a feature axis. If the straightness control were eliminated in the case of Figure 2-6, the maximum mating size would be a .502 diameter. The example in Figure 2-6 shows us that the diameter (±.002) is more critical to the design than the straightness. This may be true for parts of considerable length that may bow under their own weight, but may straighten at assembly. The use of the phrase "PERFECT FORM AT MMC NOT REQD" beneath the control frame will help ensure communication of exact design intent.

Virtual condition is calculated by the following:

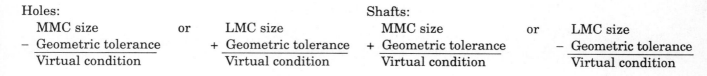

Holes:

$$\begin{array}{c} \text{MMC size} \\ - \text{Geometric tolerance} \\ \hline \text{Virtual condition} \end{array} \quad \text{or} \quad \begin{array}{c} \text{LMC size} \\ + \text{Geometric tolerance} \\ \hline \text{Virtual condition} \end{array}$$

Shafts:

$$\begin{array}{c} \text{MMC size} \\ + \text{Geometric tolerance} \\ \hline \text{Virtual condition} \end{array} \quad \text{or} \quad \begin{array}{c} \text{LMC size} \\ - \text{Geometric tolerance} \\ \hline \text{Virtual condition} \end{array}$$

(Ref. Figure 9-14)

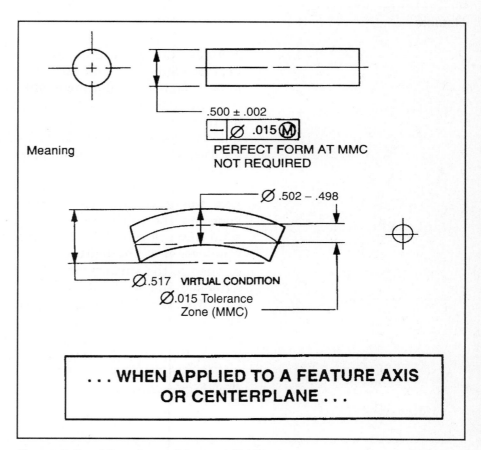

Figure 2-6. Virtual condition at MMC.

Inner/Outer Boundary

The terms *inner* and *outer boundary* are applied at RFS conditions, and refer to the collective effects of size and geometric controls for RFS.

For shafts, the outer boundary is:

Size + Geometric tolerance

For holes, the inner boundary is:

Size − Geometric tolerance

Flatness

Flatness differs from straightness in that flatness controls all surface elements in all directions (see Figure 2-7). *Flatness tolerance* specifies a tolerance zone confined by *two parallel planes* within which the *entire surface* must lie. Compared with straightness applications, flatness is fairly simple in that it can only be applied to a single, flat surface (plane). Once surfaces are verified as being flat within specifications, we may use these surfaces as *datums*.

If the design cannot tolerate abrupt steps, within the flatness tolerance, flatness per unit area may be applied. Any area zone may be selected for the incremental verification, with the total flatness value of 0.1 being applied for the entire surface. The application of the feature control frame is optional and may be applied to the surface via a leader line direct or as applied to the extension line as shown.

Waviness is a term to describe secondary surface texture. Waviness is the height and length of surface undulations which, much like the swells at sea, occur within the high and low tide limits. Waviness is contained within the flatness tolerance. See ANSI B46.1 and ANSI Y14.36 for more on texture of surfaces. (See Figure 2-8.)

Circularity (Roundness)

Circularity is a condition of a surface (of revolution) where:

1. All points of the surface intersected by any plane perpendicular to an axis are equidistant from that axis.

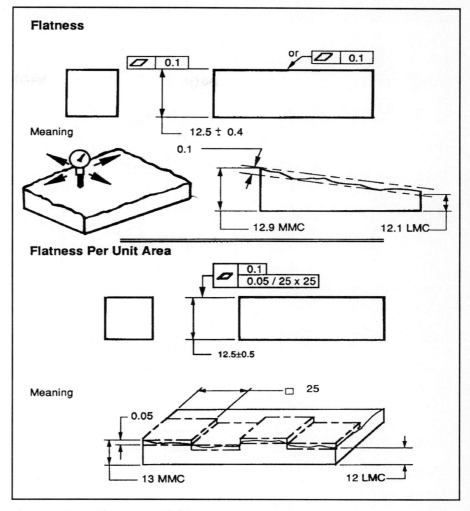

Figure 2-7. Flatness and flatness per unit area.

2. With respect to a sphere, all points of the surface intersected by any plane passing through a common center are equidistant from that center.

Circularity tolerance specifies a tolerance zone bounded by *two concentric circles* within which each circular element of the surface must lie regardless of size. Circularity is a form control for cylindrical or spherical features. Circularity is two dimensional in that it does not control depth. By definition, it is the tolerance space between two concentric circles, and thus a radial difference. (See Figure 2-9.) With circularity, feature control frame placement is not critical to design intent and may be optional.

Note that circularity error may not be detected by caliper or micrometer type measurement on shafts with an odd number of

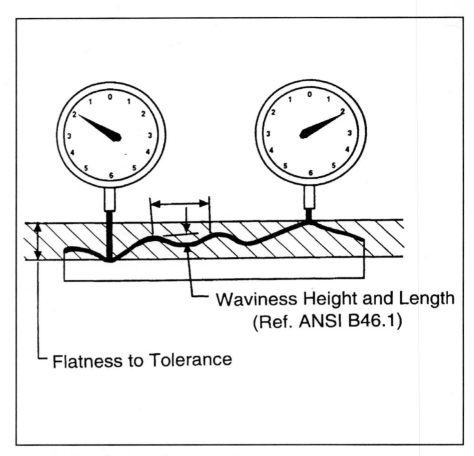

Waviness Height and Length
(Ref. ANSI B46.1)

Flatness to Tolerance

Figure 2-8.   Flatness tolerance and waviness.

lobes. This type of measurement can reflect a constant diameter measurement, but will not detect radial error. (See Figure 2-10.)

It is important to recognize that vee blocks are not always reliable in roundness measurement and assessment, as certain angles will hide the lobing. *Lobing* is most commonly found in ground shafts and at an odd number from 3 to 19 lobes. Further, if more than one supplier or plant is involved, care must be taken that all involved are using similar methods to ensure the same quality levels.

With Rule #1 in mind, review Figure 2-11. With no circularity control specified, the conditions allowed by Rule #1 will result in an MMC envelope of .860 diameter, and a cross section of .840 diameter at LMC. These diameters may be displaced, which would result in a "D" shaped section as shown. To remove this possibility, we must consider the use of circularity control as shown in Figure 2-12. It is important to note that since circularity tolerance is the space between two concentric circles within size tolerance, the circles do not have to be in alignment at various locations along the shaft.

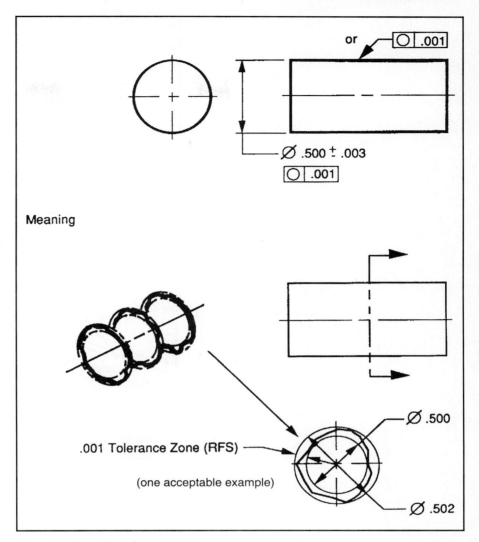

Figure 2-9. Circularity (RFS).

Circularity is the only geometric form control with an additional written standard relative to measurement practices. ANSI B89.3.1 covers circularity measurement techniques and illustrates methods of specifying circularity tolerance required for critical features when not only design tolerance is needed, but also when the precise method(s) of measurement become integral to the design specification. Not only are the tolerances given, but the method of measurement in the feature control frame is also specified (i.e., cycles or filter response and the stylus tip radius). See Section 9, Figure 9-10 and ANSI B89.3.1 for more details.

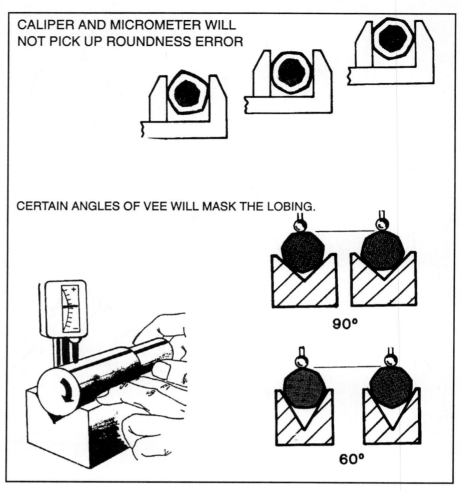

CALIPER AND MICROMETER WILL
NOT PICK UP ROUNDNESS ERROR

CERTAIN ANGLES OF VEE WILL MASK THE LOBING.

90°

60°

Figure 2-10.   Caliper micrometer and vee locator measurement error.

Cylindricity

Cylindricity is a condition of a surface of revolution in which *all points of the surface* are equidistant from a common axis. *Cylindricity tolerance* specifies a tolerance zone bounded by two concentric cylinders within which the surface must lie. Cylindricity is the control that pulls roundness and straightness controls together to form a tolerance zone made up of two concentric cylinders. This control is considered somewhat difficult to measure with conventional equipment as the tolerance zone is the space between two concentric cylinders (a radial difference). The placement of the control frame is optional but has only one meaning. (See Figure 2-13.)

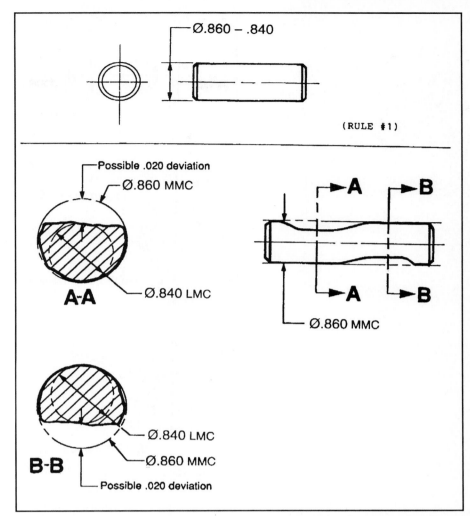

Figure 2-11. With the limits of size control only, the MMC and the LMC liimits do not have to be concentric.

Computerized measurement machines may be programmed to calculate cylindricity based on probe type measurements using a formula that is based on the square root of the sum of the roundness, straightness, and taper squared, as shown in Figure 2-14. If the measured values are all .001 as shown, the program may calculate a cylindricity value of .0017. If, however, the program only uses straightness and roundness, the calculated cylindricity would be .0014. The rounding practices in place at a plant or lab could have a great impact on acceptance or rejection of parts. It may be worthwhile to investigate the type and application of programmed cylindricity software when using these devices. (See Figure 2-14.)

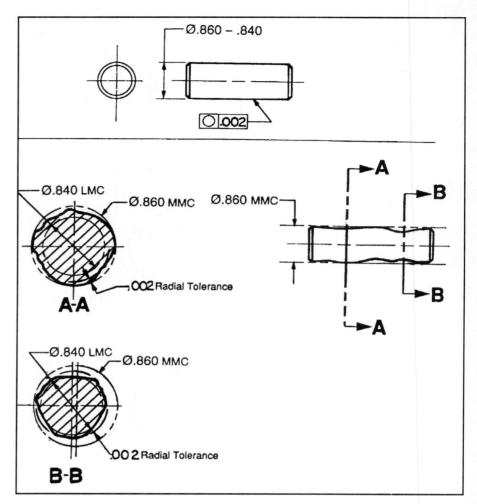

Figure 2-12.   A circularity tolerance specifies a tolerance zone bounded by two concentric circles.

Form Control Measurement

As noted previously, form controls may be more easily defined than measured. Form controls have no datums from which to set-up as they have no datum relationships. Figures 2-15 through 2-18 show examples of common measurement options when computer assistance or sophisticated equipment is not available.

Form Tolerance Summary

To understand form tolerance, you must remember that straightness (surface elements) and circularity are two dimensional con-

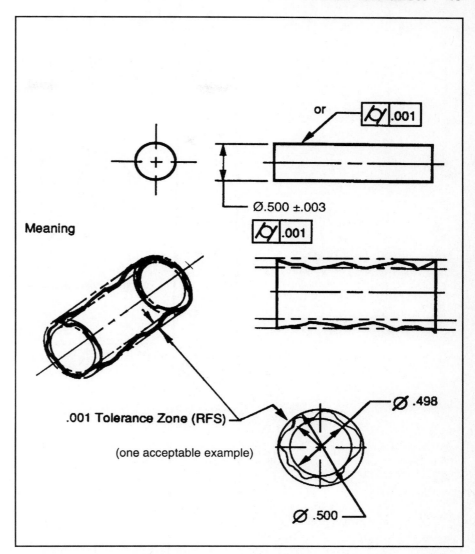

Figure 2-13.   Cylindricity (RFS).

trols, while flatness and cylindricity are three dimensional. Straightness controls surface line elements, while flatness controls entire surfaces (single direction control versus multi-direction control). Straightness control may be applied to a feature axis, using a diameter symbol, creating a three-dimensional (cylinder) tolerance zone.

The circularity controls of a circular shape apply at the surface of any given cross section normal to the feature axis, while cylindricity controls applies at the feature surface for the entire length of the feature. Circularity may be contained within the cylindricity limits.

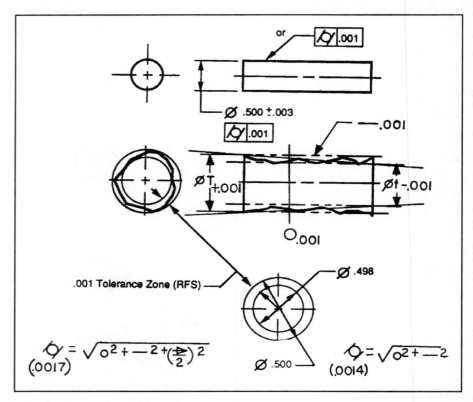

Figure 2-14.   Cylindricity measurement programs.

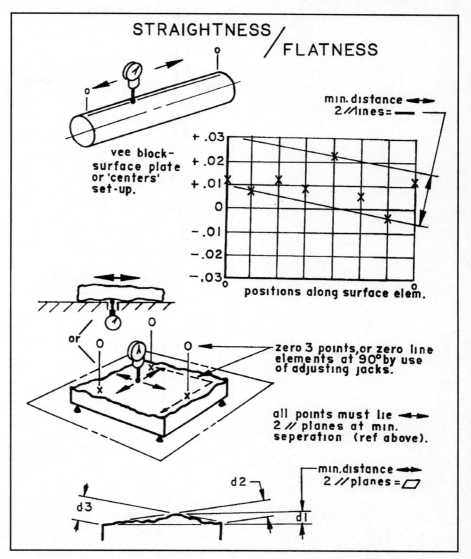

Figure 2-15.   Straightness/flatness. Measurement methods and evaluation.

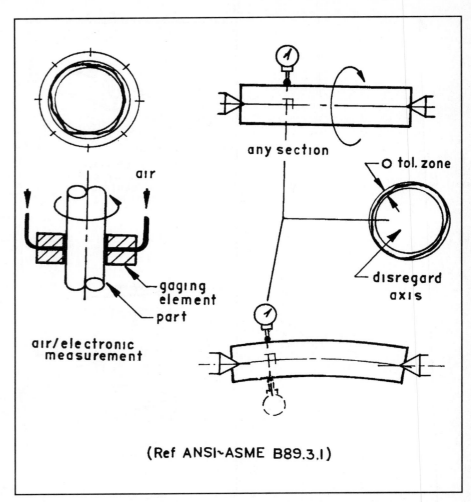

Figure 2-16.   Circularity. Evaluation methods include mechanical, optical, electronic, and air gaging. Roundness measuring machines which record the feature contour on polar graphs are quite widely used.

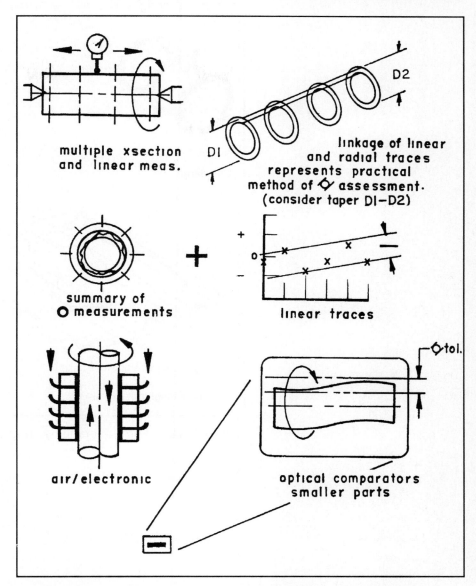

Figure 2-17.   Cylindricity. Measurement methods and evaluation.

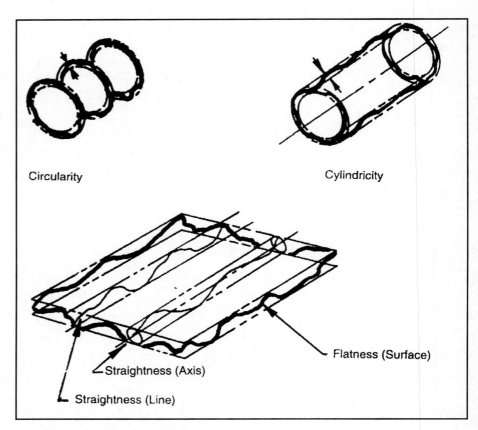

Circularity

Cylindricity

Flatness (Surface)

Straightness (Axis)

Straightness (Line)

Figure 2-18.   Form tolerance summary.

EXERCISE 2-1.   CIRCULARITY (ROUNDNESS)

1. Circularity is applied MMC unless otherwise specified.   T   F
2. Circularity tolerance controls apply at the _____ of a cylindrical shape.
3. Circularity controls *may* be applied to a tapered shaft.   T   F
4. Circularity is always datum related.   T   F
5. Circularity is a refinement of size, therefore, a form control.   T   F
6. Vee blocks are best for circularity measurement.   T   F
7. Circularity tolerance must be within size limits.   T   F
8. The use of MMC will allow a bonus circularity tolerance.   T   F
9. Indicate a circularity callout with a 0.1 tolerance in the figure below. Draw the tolerance zone.

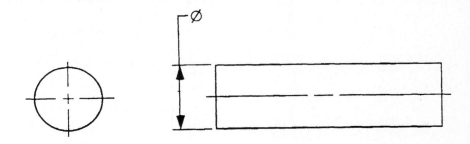

EXERCISE 2-2. CYLINDRICITY

1. Cylindricity is applied MMC unless otherwise specified.
T   F

2. Cylindricity is applied to cylindrical shapes and must be larger than size tolerances.
T   F

3. Cylindricity controls _____ and _____ of surface elements.

4. Cylindricity is always datum related.
T   F

5. Cylindricity tolerance zone is the space between two concentric circles.
T   F

6. A cylindricity tolerance may extend beyond the MMC size envelope.
T   F

7. Cylindricity may be used to control taper in the figure below.
T   F

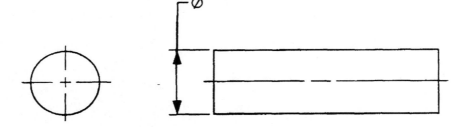

8. Indicate a cylindricity callout with a 0.2 tolerance in the figure above. Draw the tolerance zone.

EXERCISE 2-3.   FLATNESS

1. Flatness is applied MMC.   T   F
2. Flatness also controls _____ of surface elements.
3. Flatness tolerance is additive to size tolerances.   T   F
4. Flatness is always datum related.   T   F
5. If flatness controls an entire surface, then it also controls the squareness of the surface to a surface plate.   T   F
6. The figure below *may* have a flatness tolerance of 0.5.   T   F

12.5± 0.3

7. Can MMC be applied to the *surface* in the above figure?   Yes   No
8. Indicate a flatness callout (on the top surface) with a 0.2 tolerance in the figure above. Draw the tolerance zone.

EXERCISE 2-4.   STRAIGHTNESS

1. Straightness is applied RFS unless otherwise specified.   T   F
2. Straightness may be applied to surface _____, cylindrical _____, or _____.
3. Virtual condition is invoked when straightness is applied to a feature _____.
4. Straightness is always datum related.   T   F

Complete the table below for the measured sizes given. Draw the tolerance zone.

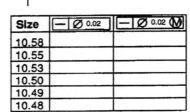

Size	— ⌀ 0.02	— ⌀ 0.02 Ⓜ
10.58		
10.55		
10.53		
10.50		
10.49		
10.48		

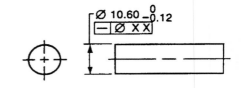

What is the virtual (mating size) condition?

Complete the table below for the measured sizes given. Draw the tolerance zone.

Size	— 0.10
14.90	
14.80	
14.70	
14.60	

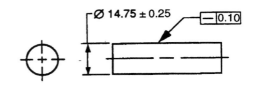

What is the virtual condition?

EXERCISE 2-5.   TRUE/FALSE

With Straightness of Surface Elements:

1. Ⓜ or Ⓛ may not be applied.
2. Perfect form at MMC is required.
3. Straightness tolerance must lie within size tolerance.
4. Virtual condition will allow Rule #1 to be violated.
5. RFS is understood to apply.

With Straightness of an Axis:

1. Datums are applied.
2. Rule #1 does not apply when straightness is applied to a feature axis or centerplane.
3. Virtual condition is equal to MMC size plus axial straightness tolerance.
4. With axial straightness, the diameter symbol is optional.
5. The mating size (virtual condition) of a shaft that is 15mm diameter ±0.3mm, with an axial straightness of 0.3 diameter is 15.3mm.

3

Datums

A *datum* is a theoretically exact point, axis, or plane derived from the true geometric counterpart of a specified datum feature. A datum is the origin from which the location or other geometric characteristics of a feature of a part is established.

A logical progression from the control of feature surfaces via form controls is the use of these features (lines, planes, and cylinders) as datum control features. But first, we must develop and understand the datum framework. Basic geometry shows us that three points will form a plane. From this plane may be constructed a second plane, perpendicular to the first, by the use of two points. A third plane may be established perpendicular to the other two by one point, thus establishing three mutually perpendicular planes or a datum framework. This three-plane concept is the foundation for defining three-dimensional objects. (See Figure 3-1.)

Datum Reference Frame

A completely defined datum reference frame will constrain or restrict a feature or part in three translational directions (x, y, and z), and in three rotational orientations (u, v, and w), where u is in rotation about the X axis, v is rotation about the Y axis, and w is rotation about the Z axis. A datum reference frame need not contain controls for all degrees of freedom for locating and orienting tolerance zones. For example, a datum reference frame defined by

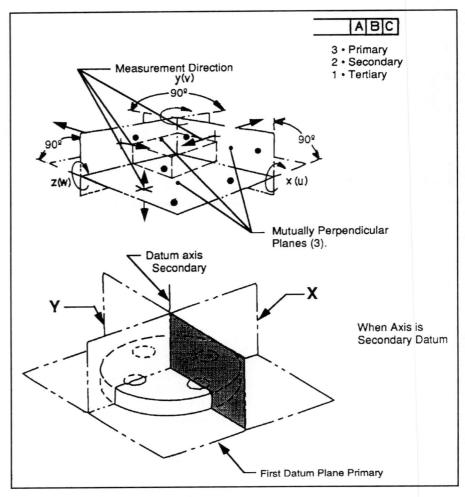

Figure 3-1. Datum reference frame.

only one datum plane will be sufficient to control perpendicularity or parallelism to that plane.

Two sets of conditions may exist with cylindrical objects or features. The first set of conditions exist with a cylindrical shape of short length where the primary design influence rests with the flat surface (primary), and with the secondary and tertiary datum planes intersecting at right angles to form the feature axis (see Figure 3-1b). The second set of conditions is where the axis of a feature of greater length, possibly with linear or rotary motion involved, exerts the greater functional influence, thus primary, with the secondary datum a potential single point stop, and the tertiary datum serving to stop any rotation in the datum reference frame (see Figure 3-2).

In Figure 3-2, three equally spaced tool elements may close to isolate the feature thus establishing the feature axis as primary design influence, while the end of the shaft stops secondary or lateral movement and the keyslot stops rotational movement, locking

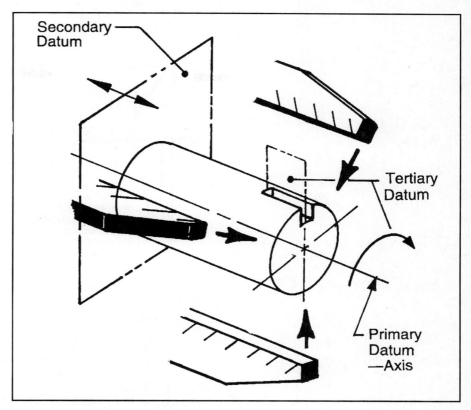

Figure 3-2. Axis—primary datum.

the part in the datum reference framework. Generally, two issues help establish when the axis might be considered the primary influence:

1. The ratio of surface area to diameter length (static applications).
2. The rotary or linear movement of the cylindrical feature (dynamic applications).

In most designs, this influence will be clear.

Datum Terms

From Figure 3-3, we can see *theoretic datums* are perfect; *simulated datums* exist in machining or measurement fixtures, equipment and simulate theory; *datum features* have process error; and *temporary datums* are for manufacturing or inspection only and may not exist on the finished product. When we talk about parts and datums, it is important to understand the terms involved. Generally, however, when referring to "datums," we tend to imply the theoretical surfaces and axes involved.

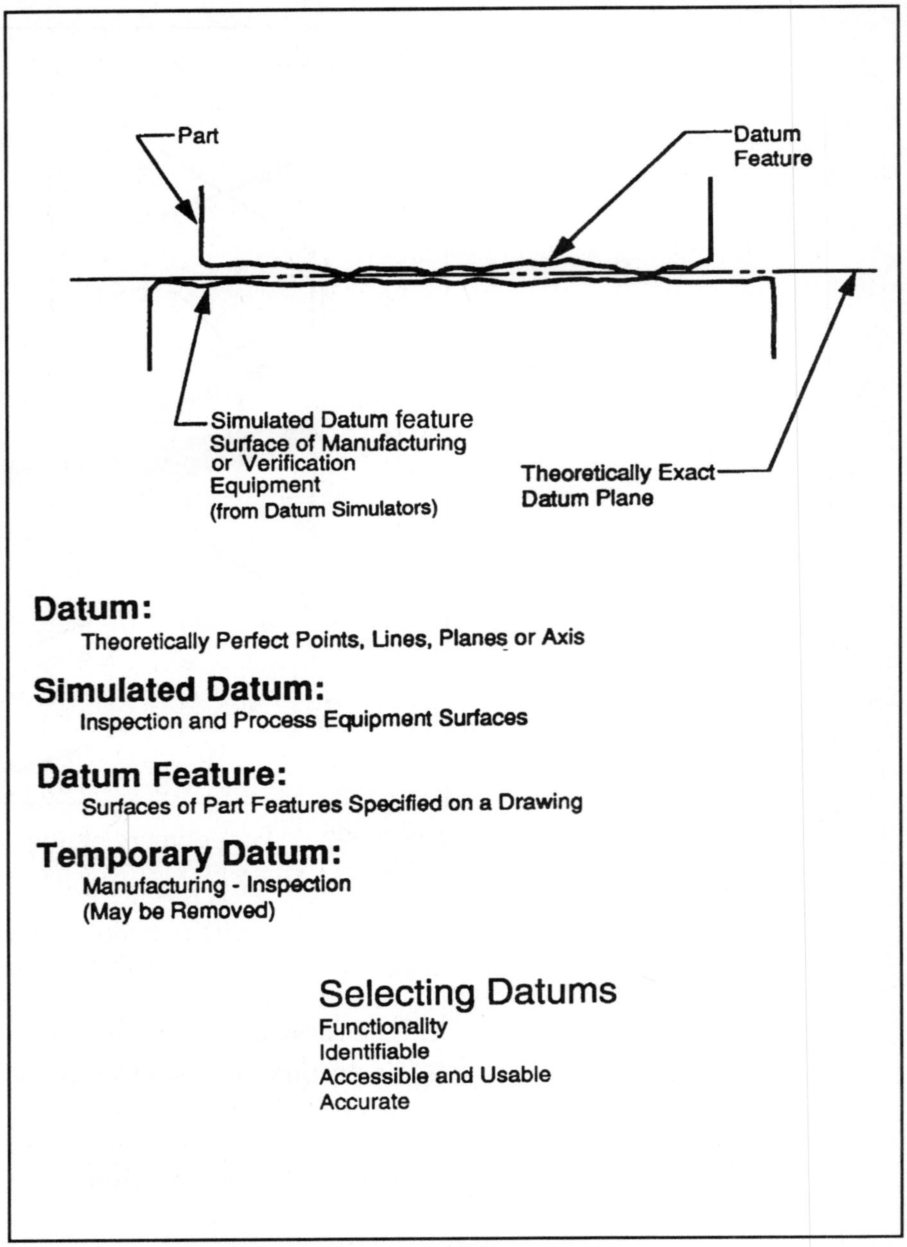

Figure 3-3.   Datum terms and selection.

Datum Selection

Once comfortable with the datum frame, theory, and terms, we are ready to select datums based on:

- *Functional relationships*—as this is a very important issue, impacting not only on the design requirements, but on manufacturing and quality plans as well.

- *Reality*—as datums should be real and identifiable. Imaginary points in space or features which are impossible or difficult to locate are not worth very much. Neither are datums which are inaccessible or hidden within the product.

- *Accuracy*—we all agree datums should be accurate as datum error can impact on other features and measurement. If a feature or surface is to be flat within a tolerance and meets that requirement, it then becomes theoretically perfect for the measurement of other features.

Remember, a specification which cannot be measured need not exist, and a datum which cannot be found (for the basis of other measurements) may be equally worthless.

Datum Feature Symbol

The datum feature symbol consists of a capital letter enclosed in a square frame with a leader line extending from the frame to the concerned feature and terminating with a triangle. The datum feature symbol (see Figure 3-4) is applied as follows:

(a) Placed on the outline of a feature surface or an extension line of the feature outline (but clearly separated from the dimension line) when the datum feature is represented by the extension line or feature surface itself.

(b) Placed on an extension of the dimension line of a size feature when the datum is the feature axis or center plane. If there is insufficient space for the two arrows, one of them may be replaced by the datum feature triangle.

(c) Placed on the outline of a cylindrical feature surface or an extension line of the feature outline, separated from the size dimension, when the datum is the axis.

(d) Placed below and attached to the feature control frame when the feature, or group of features, controlled is the datum axis or datum centerplane.

(e) Placed on the drawing planes established by datum targets on complex or irregular datum features. (See Figures 3-36, 5-10, and the section on profile controls in Chapter 5.)

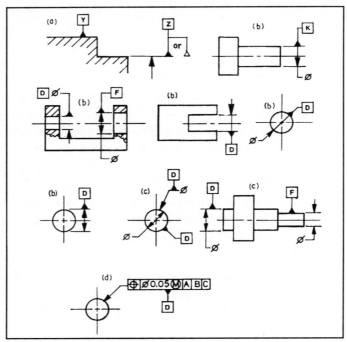

Figure 3-4a. Datum symbols and application.

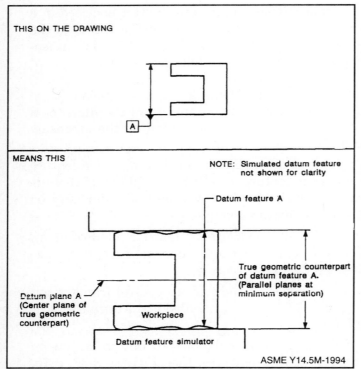

THIS ON THE DRAWING

MEANS THIS

NOTE: Simulated datum feature not shown for clarity

Datum feature A

True geometric counterpart of datum feature A. (Parallel planes at minimum separation)

Datum plane A (Center plane of true geometric counterpart)

Workpiece

Datum feature simulator

ASME Y14.5M-1994

Figure 3-4b. Primary external datum width—RFS.

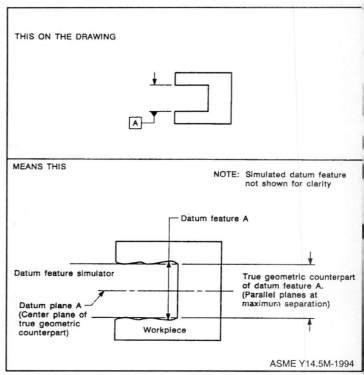

THIS ON THE DRAWING

MEANS THIS

NOTE: Simulated datum feature not shown for clarity

Datum feature A

Datum feature simulator

True geometric counterpart of datum feature A. (Parallel planes at maximum separation)

Datum plane A (Center plane of true geometric counterpart)

Workpiece

ASME Y14.5M-1994

Figure 3-4c. Primary internal datum width—RFS.

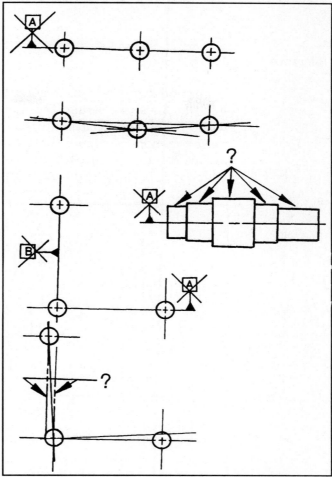

Figure 3-5. Use of centerlines (axes) for datum reference.

Datum Centerlines (Axes)

It is generally unwise to identify centerlines of features or centerlines between features as datums (see Figure 3-5). Three holes in alignment create a problem in that only two of the holes can ever be aligned. The third (any hole) will have some location error.

Hole centerlines placed at 90° also create potential datum errors in that the horizontal and vertical datums (which are mutually 90° apart by definition) will be almost impossible to determine without further definition of design intent.

Shafts with multiple features create the same dilemma in that we must define which features constitute datum feature A, for finding datum axis A.

Also, carefully consider your design before applying a control *directly* to an axis or centerplane as there may be more than one feature sharing the same centerline or centerplane. It may not be clear which feature the control was intended for (i.e., a counterbore and drilled hole or holes hidden behind the feature in question sharing the axis).

63

Features with Form Tolerance

Primary datum features generally will have a form control applied. For flat features, straightness or flatness generally will be used, while with cylindrical features, cylindricity or roundness will be most common. For more complex shapes, a profile of line elements or surfaces may be used. Figure 3-6 illustrates a controlled cylinder which is to be used as datum feature A.

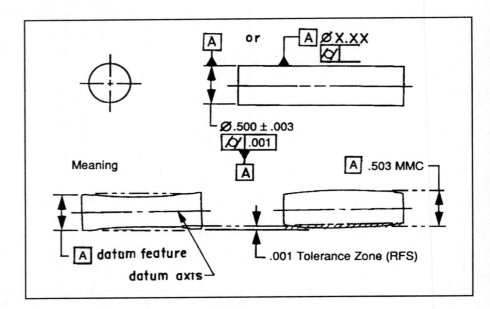

Figure 3-6.   Datum with form tolerance.

Datum Rule

When a feature of size with applied MMC is used as a secondary or tertiary datum, it must be applied at its virtual condition. (See Figure 3-7.) To determine the gaging for the secondary datum pin (datum B), we must first allow for the possibility of datum feature B being out of square to surface A. The same principle applies to tertiary datum feature slot C.

Datum Sequence

Design intent can be totally altered by changing the order of the datums, as shown in Figure 3-8. Figure 3-8a shows datum diameter A as a primary datum RFS. So we can locate on datum feature A, with perpendicularity error reflected in the secondary datum, feature B. The holes must be located primarily from and parallel to primary datum A. Figure 3-8b illustrates datum surface

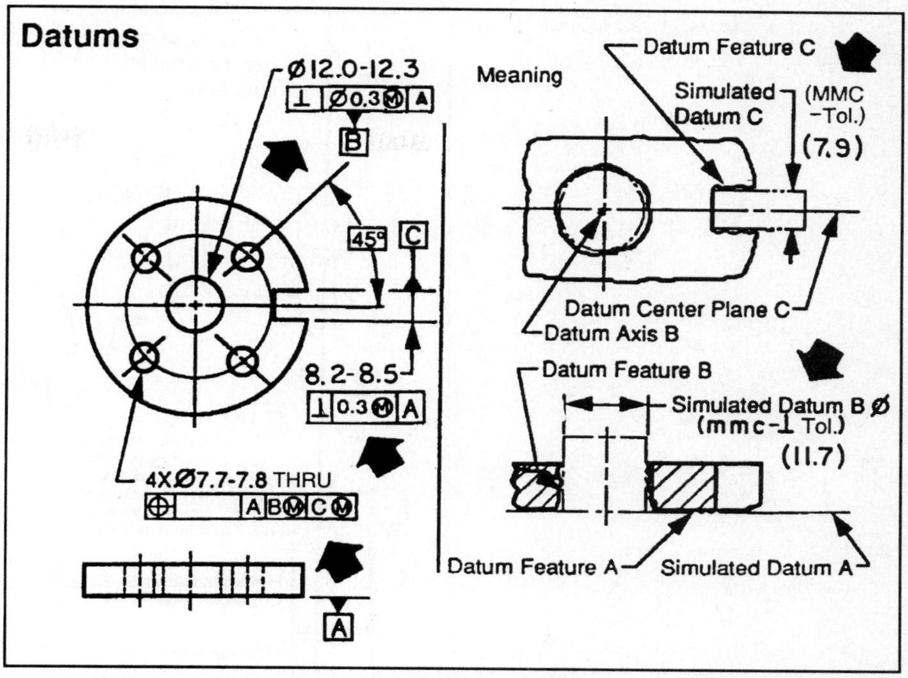

Datums

Datum Rule: Secondary and tertiary datums that are features of size require simulation at virtual condition (MMC size ± tolerance).

Figure 3-7.   Datums.

B as primary, with the error reflected as out of squareness of secondary datum diameter A. The holes must be perpendicular to primary datum B and located from secondary datum A. Secondary datum A would be the minimum cylinder contacting diameter A and perpendicular to datum B. Figure 3-8c shows surface B as the primary datum with diameter A being the secondary datum applied at MMC. This control callout will allow a functional receiver gage to be used. The gage would have a hole which simulates datum feature A at MMC (16.0) which is perpendicular to primary datum surface B. The gage surface and diameter would be mutually perpendicular within gage design tolerances. From this example, we see that it is clear that the order of datum precedence is as important as the identification of the datum features themselves.

Datum Framework

Figure 3-9 illustrates examples of datum frameworks, along with possible gage designs which have been developed from functional relationships and control callouts specified. Note when the secondary datum is a feature axis RFS, the tertiary datum serves to orient the part (stop rotation). (See Figure 3-9b.)

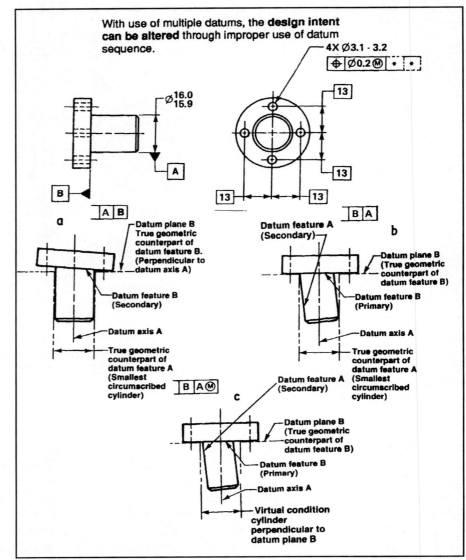

Figure 3-8.   Datum sequence/precedence.

(ASME Y14.5M-1994)

Secondary and Tertiary Datums

Tertiary Datums

If the first two datums leave a rotational degree of freedom, the tertiary datum will be *basically oriented but not necessarily basically located* to datums of higher precedence. (See Figure 3-10.)

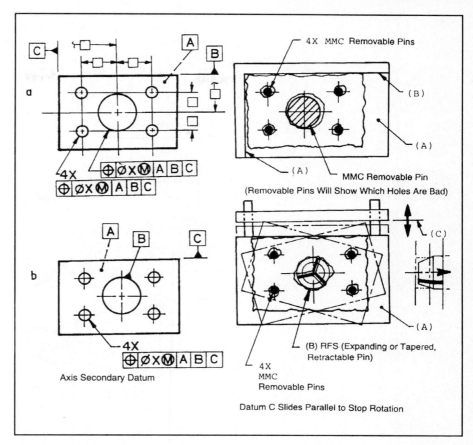

Figure 3-9.   Datum framework.

Datum Features of Size (RFS)

Cylinder/Width. The axis or centerplane of the RFS actual mating envelope of the datum feature constrained is to be *basically located and oriented* to the higher precedence datums.

Sphere. The centerpoints of the actual mating envelope of the datum feature are *basically located* to the higher precedence datums.

Datum Features of Size (MMC/LMC)

Cylinder/Width. The axis or centerplane of the virtual condition size envelope of the datum feature constrained is to be *basically located and oriented* to the higher precedence datums.

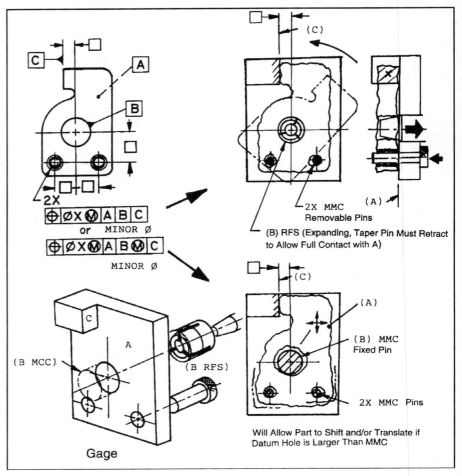

Figure 3-10.   Datum framework (axis secondary datum).

Sphere. The centerpoints of the virtual condition size of the datum feature are located from the higher precedence datums. (See Figures 3-9, 3-10, and 3-11.)

When datum features are cylindrical, the datum framework may be established by 2 feature axes as shown in Figure 3-11.

Primary datum axis A will establish the intersection of two planes (A & B), while the tertiary datum C is established perpendicular to A & B by the axis of the small hole datum B. This concept stops all rotational and lateral movement of the part in the framework, thus locking the part in the gage fixture.

Note with cylindrical parts, when the primary datum is a feature axis, and the secondary datum is also a feature axis, not in the same plane as the primary datum feature, only two datums will establish the datum framework. This is because cylindrical features are features of size, with an axis (or center) and combinations of the two planes for each datum feature will allow development of the three mutually perpendicular planes.

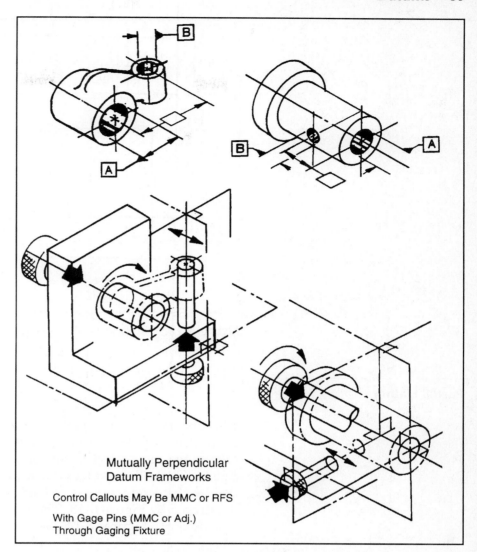

Figure 3-11.   Datum framework (two axes).

Stepped Datum Feature

Three points establish a plane. Figure 3-12 illustrates how we can offset one of these points by a basic amount to establish datum plane A. This condition is common to many parts and is referred to as a *stepped datum*. The 25mm basic dimension is for the basic separation and tool/gage tolerance application, while the 25 ± 0.2 dimension is for part tolerance. Note also the use of the dimension origin symbol indicating the precise tolerance zone location.

Target point symbols also have been introduced to identify the three locations for establishing datum A. We will cover datum targets in more detail later in this section.

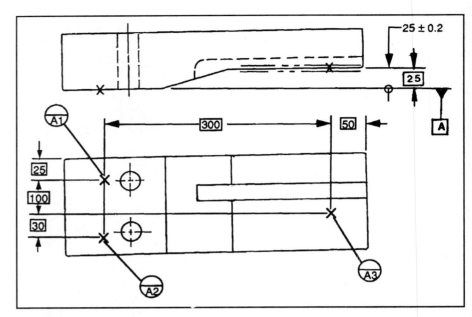

Figure 3-12.   Stepped datum feature.

Partial Datum

Often we encounter part features which are very long or otherwise inappropriate to use in their entirety as a datum. When this occurs, we may select a portion of the feature to use as a *partial datum*. These limited lengths or areas may be identified and specified by the combined use of basic dimensions and target points as illustrated by Figure 3-13. The use of datum symbols is also acceptable. Refer to Figures 3-14, 3-30, and 3-31.

Coplanar and Coaxial Datums

When two part features are used simultaneously to establish a single datum reference (axis or surface), the condition is referred to as a *coplanar* or *coaxial datum* (see Figure 3-14). In this case, the datum symbols appear in the same box of the feature control frame, but are separated by a dash (A-B). This callout indicates that two features are used as one. Further, it is acceptable to use combinations of datum symbol callouts and datum target symbols. In the case of a shaft, we may wish to use the entire diameter of the shaft or select specific spots of the shaft located with basic dimensions and identified with datum target symbols. Now the datum system is closely

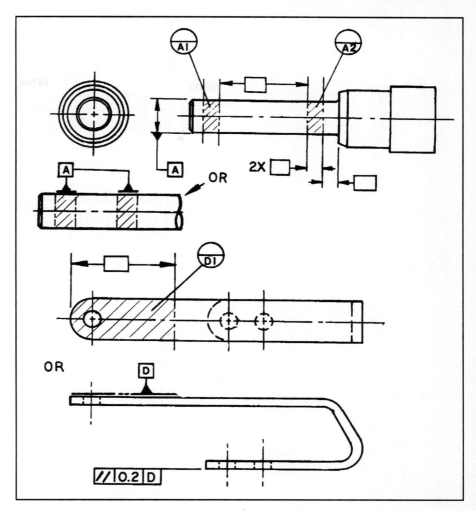

Figure 3-13.   Partial datum.

defined. We might choose to use this method on parts where precise repeatability is critical and/or when we have multiple plants or suppliers involved as the measurement results may vary depending on set-up methods used. (See the enlarged view of Figure 3-14.)

Hole Patterns As Datums

Not all parts are rigid, stable, or designed in such a way that datums are easily identified or practical to use. Gaskets, shim stock, moldings, or stampings are good examples. When no other functional datums are available, we may consider mounting holes (as a pattern and as a possible option). Figure 3-15 illustrates a part with many holes. This type of part may

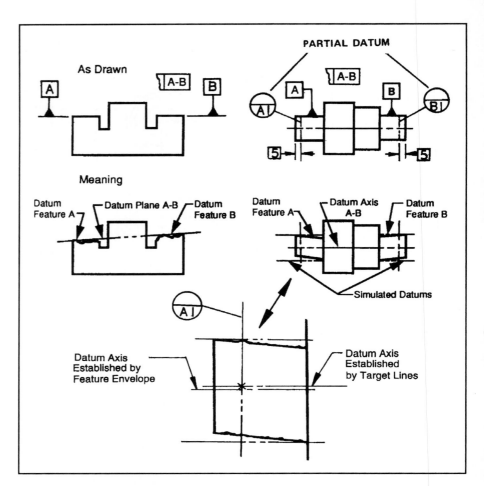

Figure 3-14. Coplanar and coaxial datums.

be a large cover plate or sump type pan. By using the hole pattern as our secondary datum part MMC, we can use a functional receiver gage as an option for part acceptance. Start with gage pins at MMC (minus tolerance), to accept the part, and evaluate the inner and outer contour for conformance to the profile control limits. This technique is popular on high-volume parts where part acceptance is the primary concern. Naturally, basic dimensions would be required for the hole locations as well as the contours.

Hole patterns may also be used as datums if it is necessary to locate another feature (hole) centered within the pattern as shown by Figure 3-16. This dimensioning technique ensures the pattern is centered in the part (symmetrically from feature centerplanes), and that the large hole is centered within the pattern of holes when both the holes (datum) and feature (large hole) are at MMC.

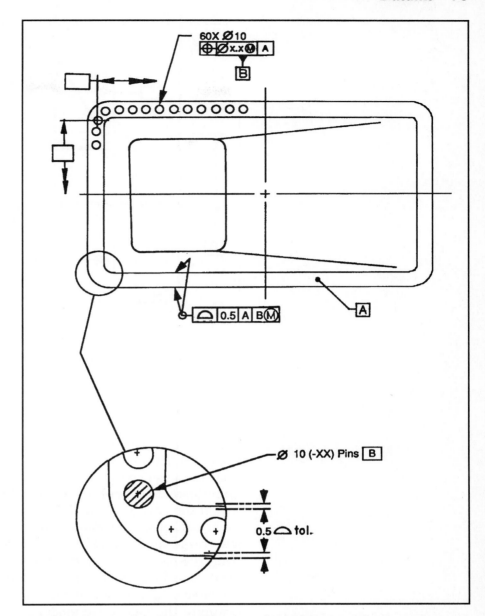

Figure 3-15.   Hole pattern datum. Multiple features of size, such as a pattern of holes at MMC, may be used as a group to establish a datum when part function dictates.

Feature Centerplanes

Determining feature centerplanes can be difficult if the parts are not square. Figure 3-17 illustrates different aspects of imperfect parts using the principle of centerplanes as datums. Figure 3-17 shows we can construct sets of parallel planes from the extremes of part edges. These planes are mutually 90° by definition. The cen-

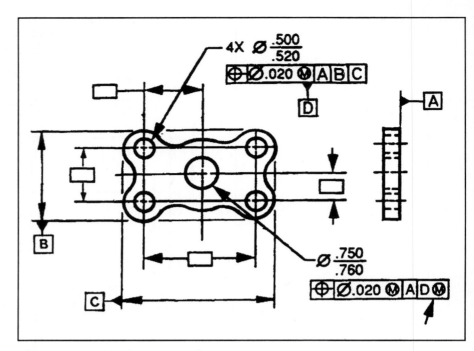

Figure 3-16. Hole pattern datums.

terplanes of this constructed geometry are the datum centerplanes B and C. If we applied MMC, we could build a functional receiver gage 6.050 × 4.050 in size, with the four-hole pattern gage pins located from the gage centerplanes. If we applied RFS, we could measure from the centerplanes of the geometry constructed RFS.

If the part was found to be skew as shown in Figure 3-17c, we could establish the maximum size of the secondary datum B at one end of the part and, by use of shims or mathematically equalizing the other end, we could construct the mutually perpendicular datum framework. Other techniques are possible as well—the use of coordinate measuring machines (CMMs) to mathematically simulate the geometric model is one. The use of profilometers or shadow graphs are another.

Datum Frameworks—Two Holes

Datum reference frames are often established by a surface (primary) with two holes (secondary and/or tertiary) as the datum frame (see Figure 3-18). Holes are features of size. Therefore, when using holes as secondary and tertiary datums, we must consider the use of RFS or MMC in the tolerance controls. Figure 3-19 illustrates possible gages with datums at both RFS or MMC. The feature holes are at MMC. Further, when using CMMs or other computer-assisted methods of measurement, we must ensure that we accept the same level of

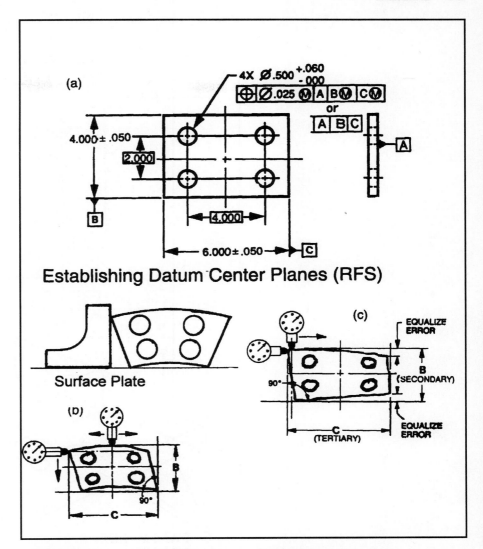

Figure 3-17. Position application and symmetry control.

quality as with a functional receiver gage. We review more positional tolerancing and gaging exercises on pages 137-201.

Gage Design Tolerances

Before we go too far, it may be well to remember that the gage is allowed a percentage of the print tolerance (ANSI B4.4). This is 5% for new gages and 5% for gage wear for a total of 10%. This tolerance normally is used by the gage designer, as required, by applying half the allowed tolerance to size of gage features and half to gage feature locations and form. Refer to ANSI B4.4 for complete information. The gage pins in Figure 3-19 illustrate this tolerance. Other examples throughout this workbook imply this tolerance, even if they are not specifically noted.

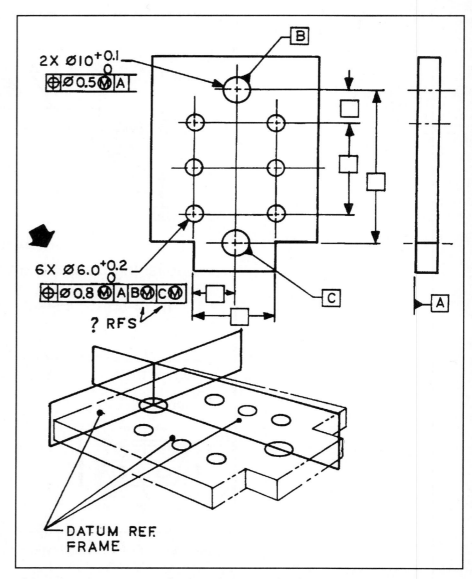

Figure 3-18. Datum reference framework—through two holes.

Datum and Feature Controls

The following figures illustrate the effects of material condition modifiers on features and datums. Study each example and determine the correlation of design controls and gages.

Figure 3-20   Datum holes at MMC are located from the framework A, F, D. Feature holes (6) at MMC are located from A, B at MMC, and C at MMC. (See Figure 3-21.)

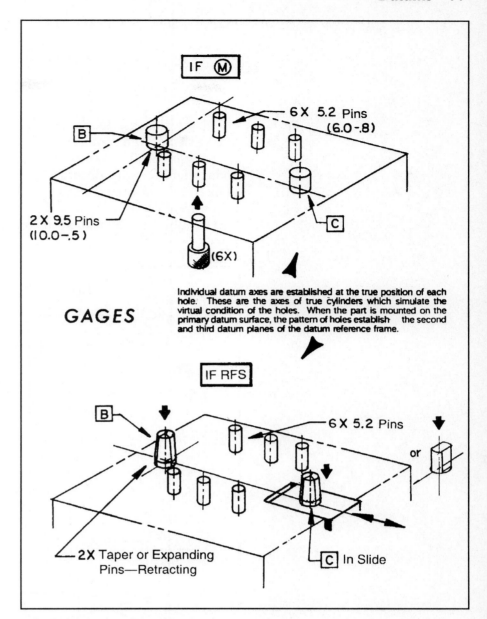

Individual datum axes are established at the true position of each hole. These are the axes of true cylinders which simulate the virtual condition of the holes. When the part is mounted on the primary datum surface, the pattern of holes establish the second and third datum planes of the datum reference frame.

Figure 3-19. Gages.

Figure 3-22	Shows the added requirement of datum holes MMC perpendicular to surface A. (See Figure 3-24a and b.)
Figure 3-23	The datum holes are located from the framework within zero tolerance at MMC. (See Figure 3-24c.)
Figure 3-25	The feature holes are located from datum holes at RFS with the feature holes at MMC. (See Figure 3-26a.)
Figure 3-25	The datum holes at RFS are located from the part datum framework A, F, D. (See Figure 3-26a.)

Figure 3-26b   The datum holes B are of equal importance at MMC, located from the datum framework A, F, D. (See Figure 3-26c.) Secondary datum axis B for the small hole location is established at the centerplanes through both datum holes B. (See Figure 3-26c.)

Note that as the datums and modifying controls change, new and varied design requirements emerge.

Figure 3-20 shows that the two 10mm holes are to be the secondary and tertiary datums. The callout shows that both holes are to be in true position relative to the datum framework A, F, D within a tolerance of 0.5 diameter when the holes are at MMC. Figure 3-21 graphically illustrates a possible functional gage for this control.

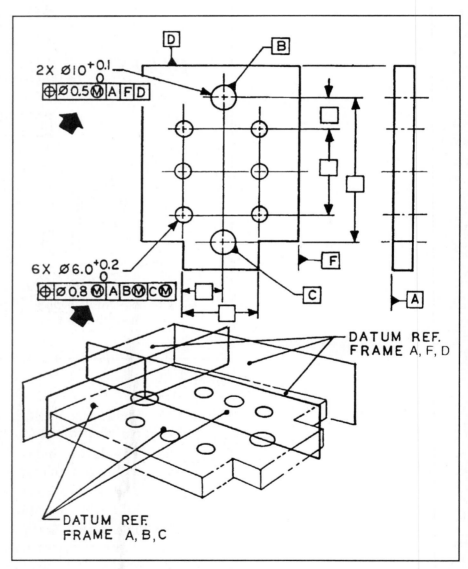

Figure 3-20.   Datum reference framework—through two holes.

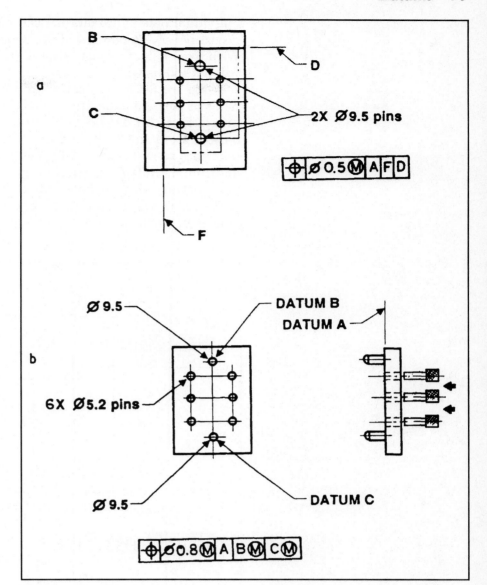

Figure 3-21.   Explanation for Figure 3-20.

Figure 3-20 also illustrates that six 6mm feature holes are to be located from surface A (primary) and that two 10mm datum holes (secondary) when the two datum holes are at MMC (10.0 mm) and when the feature holes are at MMC (6.0mm). The tolerance for the six feature holes is 0.8 diameter. Figure 3-21b illustrates a possible functional gage for this callout. A functional gage is possible here because the feature holes and the datum holes are both expressed at MMC.

Note in Figure 3-21b that six loose pins have been used for the feature holes. Fixed pins could have been used, however, with the part placed on the gage having full contact on surface A, aligned on datums B and C, the use of loose pins will determine which, if any, individual hole(s) are out of location. This gives us insight into any necessary equipment rework or repair, or drill spindle replacement. With fixed pins, the gage will tell us if the part is good or bad. But it will not tell us why.

Figure 3-22 illustrates an additional control for the datum holes of perpendicularity to surface A within a 0.2 diameter tolerance when the datum holes are at MMC size, as well as being located from the A, F, D datum framework within a 0.5 diameter tolerance zone. Figure 3-24a and 24b illustrates the impact of this additional control on possible gage designs. Both the secondary and tertiary datum holes must be perpendicular to surface A within a 0.2 diameter tolerance zone (9.8 diameter gage pins). However, as the secondary gage pin has not allowed for any posi-

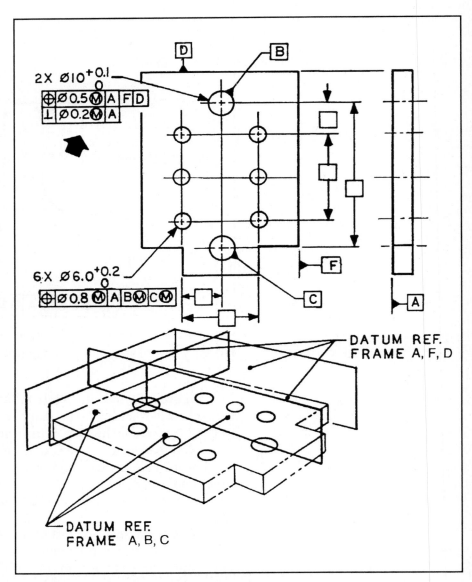

Figure 3-22.   Datum reference framework through two holes.

tional tolerance, this tolerance (0.5 diameter) is reflected into the tertiary datum. To calculate the compensation for any positional error between the datum holes, see Figure 3-24a.

Since each horizontal entry in a feature control frame is a self-contained, separate specification, it may be verified independently of other specifications or controls. Figure 3-24b illustrates the use of two 9.8 diameter pins for the secondary and tertiary datum holes to verify perpendicularity within 0.2. Because there may also be positional error between the datum holes, the tertiary datum pin is placed in a movable slide to allow for the 0.5 possible positional error. The six 5.2 feature hole pins could be added to this gage as well, giving us an optional gage design.

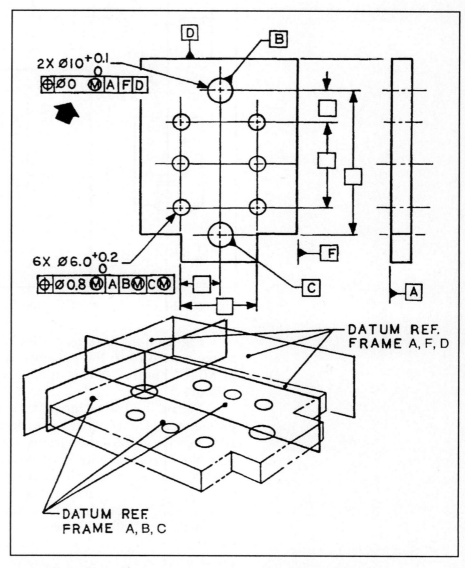

Figure 3-23.   Datum reference frame through two holes.

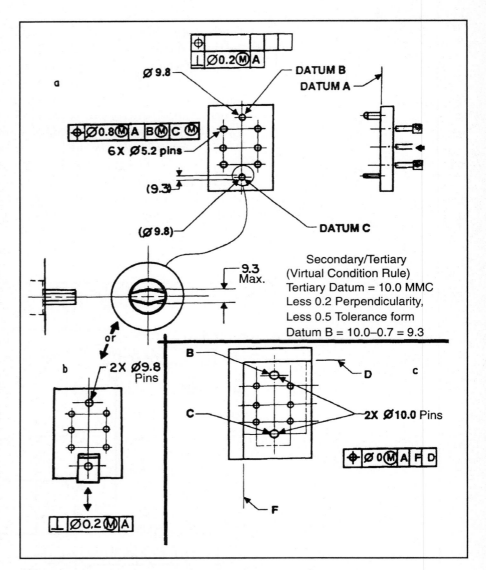

Figure 3-24.   Explanation for Figures 3-22 and 3-23.

Figure 3-23 illustrates that the two 10mm datum holes are to be located from the datum framework within a zero tolerance at MMC. Figure 3-24c graphically shows a possible gage for this callout.

Figure 3-25 illustrates two datum holes located within a 0.5 diameter tolerance zone RFS. Because the tolerance is expressed RFS relative to the datum framework A, F, D, a functional gage is not possible. Measurements must be made via surface place inspection, CMMs, or other variable measurement systems. (See Figure 3-26a.)

Figure 3-25 also illustrates the six 6mm feature holes are to be located from surface A and the two RFS datum holes with the feature holes at MMC. This callout permits six 5.2 diameter gage pins for the MMC feature holes. However, because the two datum holes are expressed RFS, some variables must be built into the gage to

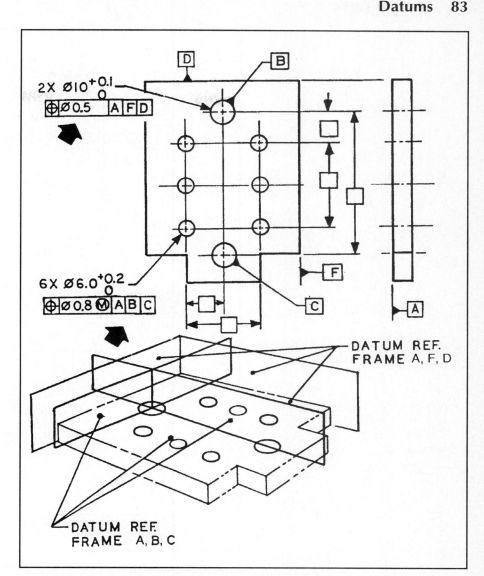

Figure 3-25. Datum reference framework through two holes.

allow for possible tolerance variations which could apply to the datum holes. Figure 3-26a illustrates the technique of using tapered, retractable pins, allowing the part to make full contact with primary datum surface A, aligning on secondary datum hole B and tertiary datum hole C. Because of possible location tolerance variations between the datum holes, the pin for tertiary datum C must be placed in a *movable slide*. Note while this approach may come close to verifying the location of the six feature holes to the datum framework A, B, C, it is not exact in that it does not verify the squareness of datums B and C to surface A. The tapered pins only contact the datum holes at surface A, not through the full depth of the datum holes.

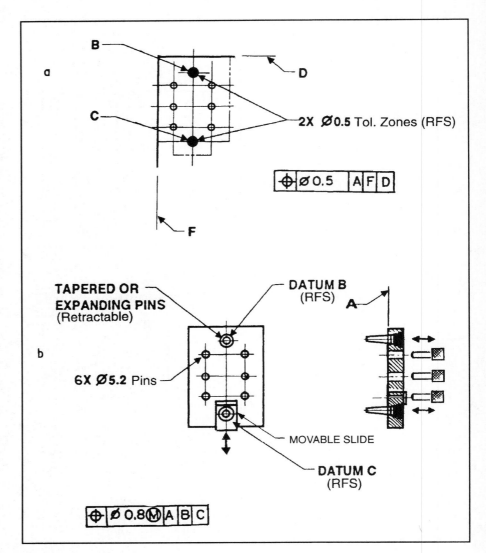

Figure 3-26a.   Explanation for Figure 3-25.

Figure 3-26b illustrates the two datum holes (as a pair) as secondary datum B (MMC). No tertiary datum is expressed. This condition is further illustrated in Figure 3-26c. Neither datum hole is shown as having more design influence.

Figure 3-26c illustrates the effect of this set of conditions with the six small holes located from surface A and the two datum holes at virtual condition MMC. The axis of secondary datum B goes through both holes, no matter which is first, therefore, the three plane framework may be established without a tertiary datum expressed. The tertiary datum plane may be constructed through either hole. This method of expression is common throughout industry. The datum holes may have been shown in the control frame RFS, however, adjustable/variable set-up methods as shown in Figure 3-26a would be required.

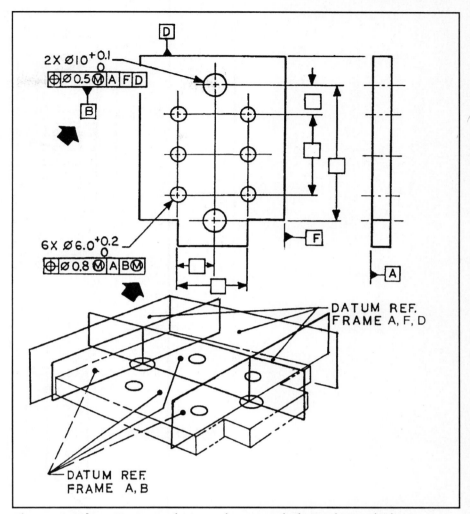

Figure 3-26b. Datum reference framework through two holes.

Datum Targets

Datum targets are often used to specifically control points, lines, or areas of surfaces which are to be used to further clarify design requirements. Targets may be used on rough, produced surfaces such as cast or forged surfaces or on finished surfaces. We use target points or lines on rough surfaces to qualify parts in order to ensure adequate machine stock, provide uniform contact points or areas for initial operations, minimize scrap, help ensure repeatability in the gaging process, and to properly communicate with suppliers. Target points are used on finished surfaces to ensure

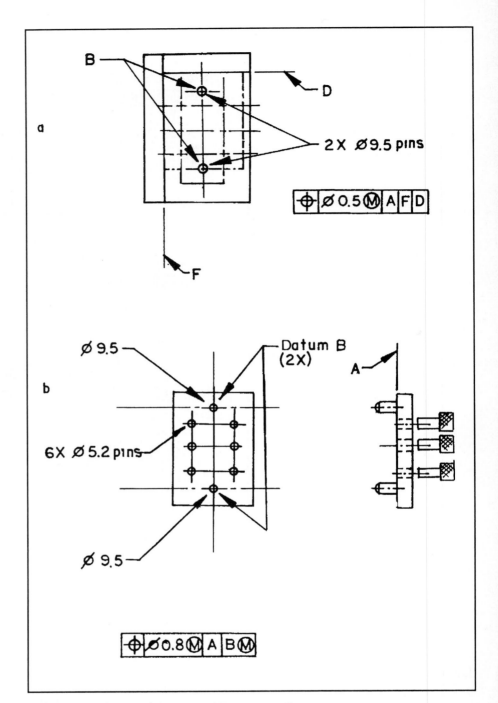

Figure 3-26c.   Explanation of Figure 3-26b.

precise controls which maintain measurement repeatability, ensure interchangeability, and gage commonality due to multiple plant or supplier sourcing. We must remember, however, that more controls mean less manufacturing and gaging flexibility. (See Figure 3-27.) Application of target points is shown in Figures 3-28 through 3-33.

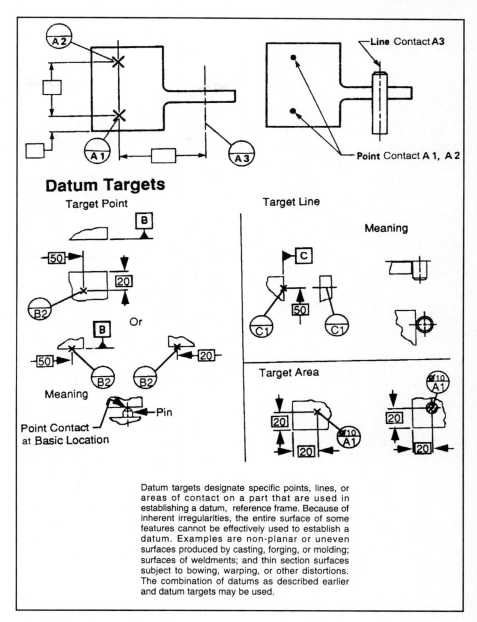

Figure 3-27.   Datum targets.

Target Points/Simulated Datums

Figure 3-29 illustrates a part which has been isolated in a fixture through the use of vee locators and target points. This practice is common when you wish to equalize casting, forging, or molding errors so holes or other features are centered in the rough part. The plane created by this method of fixturing more appropriately exists in the fixture rather than the part. The term used to identify this

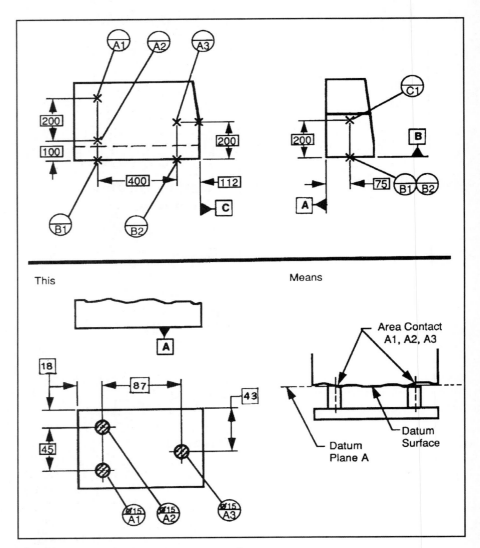

Figure 3-28.   Datum target examples.

condition is *simulated datum*. Simulated datums are appropriate when you desire to center the part in order to help equalize production errors or when feature datums are impractical or unavailable. Figure 3-29 shows a part secured in a fixture with production error equally disposed about the centerplane B.

Note that the datum target leader lines are hidden lines in the top view of Figure 3-29. This is because the target point is on the far side of the illustrated target leader. If vee locators were fixed at the left side and adjustable on the right, the tertiary datum could be established on the left end of the fixture as shown in Figure 3-30. If both the right and left fixtures were adjustable, the simulated datum could exist at the theoretical center of the fixture as shown in Figure 3-31.

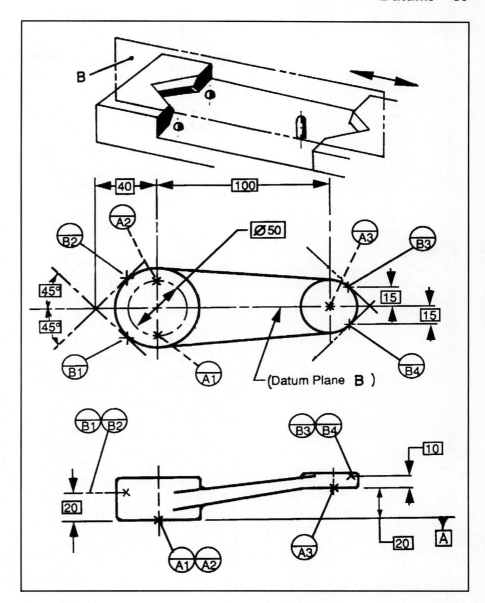

Figure 3-29.   Simulated datum plane—inspection or process equipment.

Datum Axis Primary

Recall our discussion on datum axes and target points. Figure 3-32 illustrates the use of target points to create the primary datum axis. From the diameter length to surface ratio we see that the greater influence on function appears to rest with the diameter. Due to the length of primary datum diameter A, it may be found to be impractical (or not possible) to locate on the entire diameter for evaluation or measurement. In this case, we could

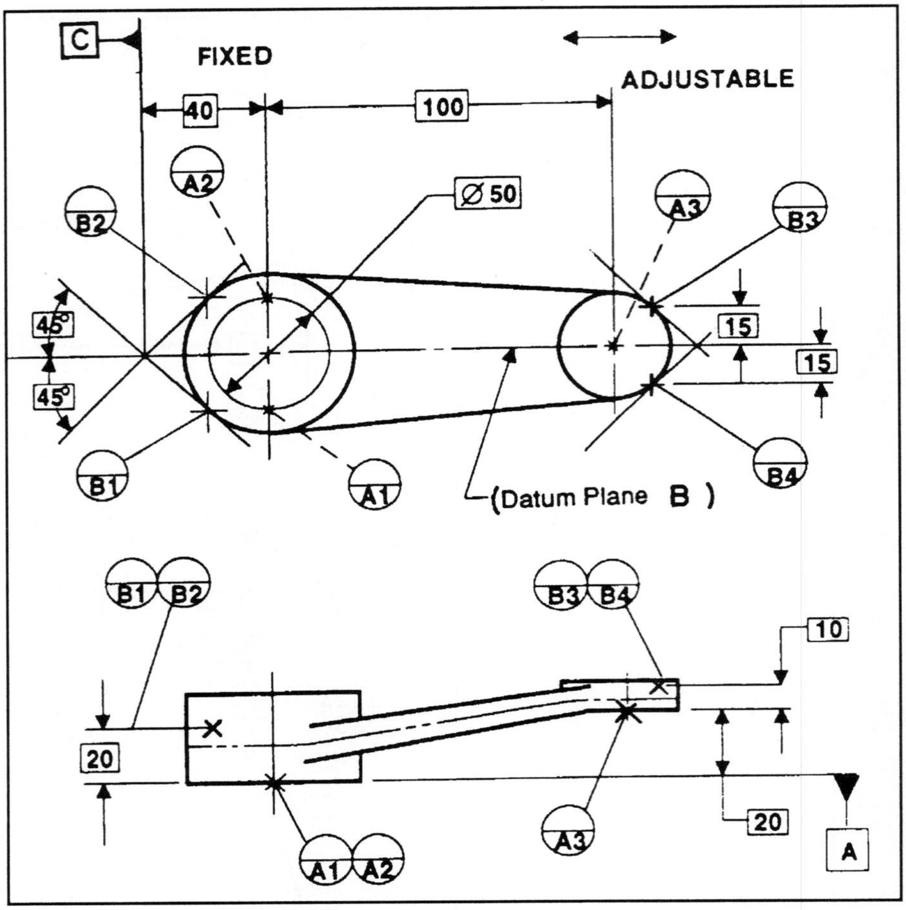

Figure 3-30.   Target points/datum planes (*cont.*).

indicate target points at each end of diameter A to aid in establishing datum axis A. (See Figure 3-32a.) In this example, we have created a partial datum, using only a part of datum feature diameter A.

In Figure 3-32b we've used the same technique. However, as diameters A and B were not of the same size (two different features), the target system creates datum axis A-B. The end surface on the shaft in both cases is the secondary datum.

Equalizing Error-Datum Targets

Sometimes the very nature and design of a part makes producing a quality product very difficult. An example might be long, thin wall stampings, moldings, or castings where production errors tend to accumulate. In some cases, industry standards refer to tolerances per

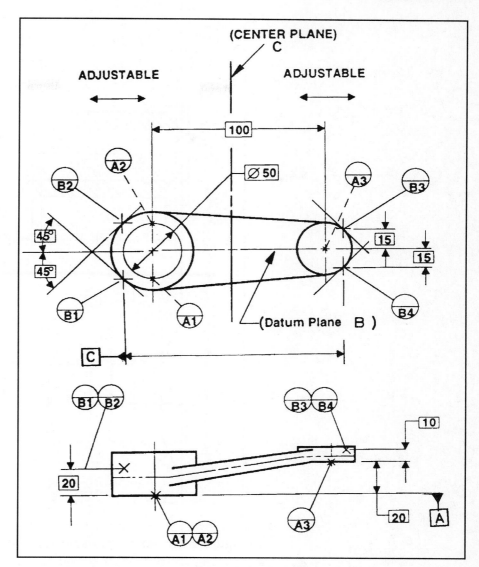

Figure 3-31. Target points/datum planes (*cont.*).

unit length, allowing errors to compound or grow according to length. One method for coping with this problem is to simulate datums using fixture centerplanes to equalize the error. The example in Figure 3-33 shows a long cover, possibly a casting, with datum targets in both the *X* and *Y* direction. These targets may be mechanical or hydraulic locators. They will actuate in the fixture, positioning the part so that when dimensioned and located from the centerplanes, production error is halved. The datum framework is mutually 90° and controlled in the fixture. Another method is to use a datum dimensioning technique to control symmetry, similar to that shown in Figure 7-13.

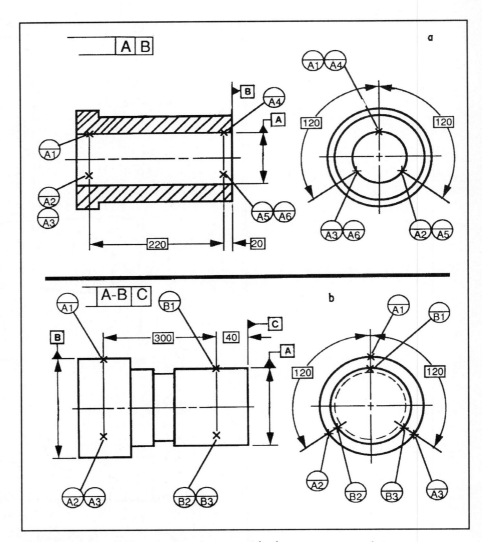

Figure 3-32.   Datum axis primary with datum target points.

Inclined Datums

Figure 3-34 illustrates a condition where the tertiary datum is established as mutually perpendicular to the other two datums, but due to the shape of the part, it is necessary to rotate it about the intersecting planes in order to locate the true geometric counterpart of surface C. The fixture would locate the part on primary datum surface A and secondary datum surface B, with the part sliding to be locked in on angular tertiary datum surface C.

Multiple Datum Frameworks

We undoubtedly will encounter designs with requirements relating to more than one set of datum frameworks. We must separate the

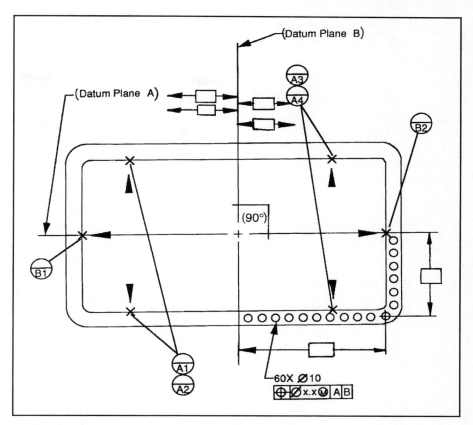

Figure 3-33.   Datum targets—simulated datums.

requirements and think through the most critical design requirements, individually, first to last, and in order of importance.

It is important to remember that in each different framework (datums in different order or different datums) constitutes a separate functional and gaging requirement. One gage will not satisfy all the requirements of Figure 3-35.

Mathematically Defined or Complex Surfaces

In some cases, we must deal with features which do not have any flat, true surfaces (airfoils and turbine blades are good examples). They nevertheless require a method of defining a datum framework for design and evaluation. The mathematically defined surface illustrated in Figure 3-36 is an example of a warp surface located in a datum framework. Points located on the surface with basic dimensions, could then be related to the framework with plus and minus dimensions, or with basic dimensions and profile tolerances. Other techniques are possible. (See also Figure 5-10.)

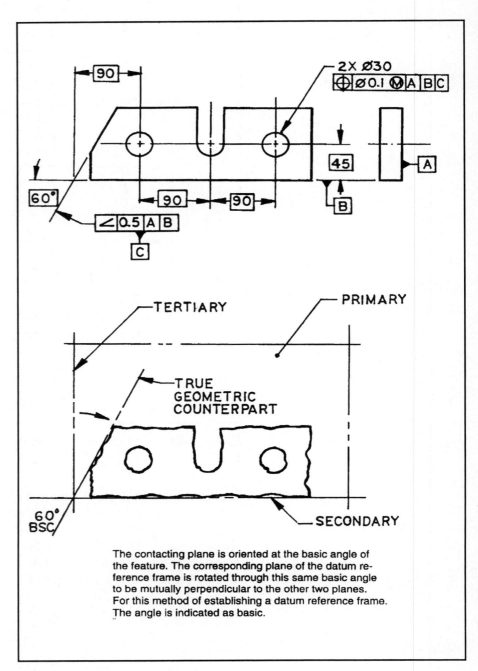

Figure 3-34.   Inclined datum features.

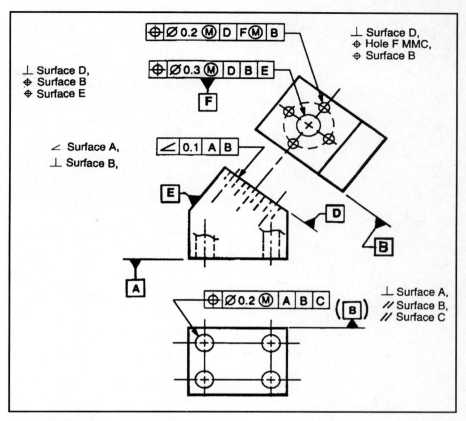

Figure 3-35. Multiple frameworks.

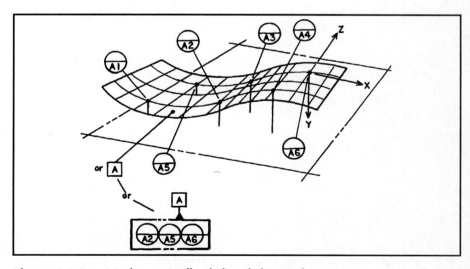

Figure 3-36. Mathematically defined datum features.

EXERCISE 3-1.   DATUMS

1. In order of importance datums are: _____,
_____, or _____.

2. Occasionally, datum features are not possible or practical to use. In this case, _____ datums are defined.

3. Ideally, datums should be selected and specified as determined by _____.

4. The figure below represents a _____ datum.

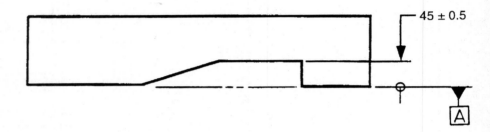

5. The datum which "stops" rotation of a cylindrical part is normally the _____ datum.

6. The use of opposite ends of a shaft simultaneously for datum purposes is shown by this callout:

✈	.005	

7. The datum framework consists of _____ mutually _____ planes.

8. In a functional receiver gage (MMC), secondary and tertiary datums that are features of size are applied at their _____

In the figure to the right:

- Specify the mounting surface in the right view as datum A.
- Specify the flat surface at the bottom as datum B.

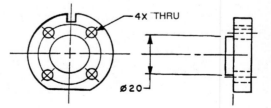

- Specify the ⌀ 20 as datum C
- Specify the slot centerplane as datum D.
- Which datum features above are subject to size variation and virtual condition? _____
- Of datums A, C, and D, what sequence and modifier symbols appear the most logical, if functional gaging is to be used? Why?

True–False

1. Datum accuracy and accessibility are equally important.
2. Datum order is determined by functional influence.
3. The datum reference frame provides stability for measurement.
4. Datum planes have length, width, and depth.
5. Actual part surfaces are called datum features.
6. Datum symbols should be indicated to centerlines and centerplanes.

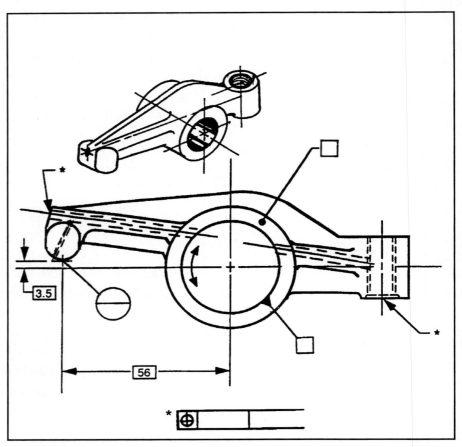

This lever operates on a shaft, with adjacent levers, to actuate movement of intake and exhaust valves.

In your judgment (using the figure above):

- What is the primary datum? **(A)**
- What are the secondary and tertiary datums? **(B, C)**
- How would you complete the control callout for the oil hole and adjusting screw hole if functional gaging is to be used?
- Show the secondary datum to be perpendicular to the primary datum within 0.5 RFS.

4

Orientation Tolerances

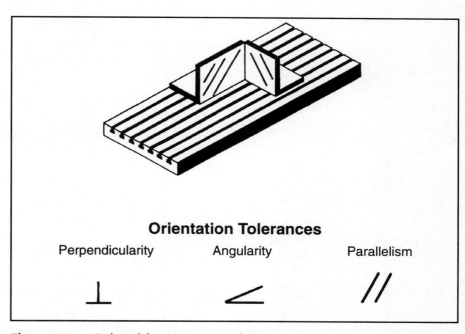

Orientation Tolerances

| Perpendicularity | Angularity | Parallelism |

Figure 4-1. Related features using datums.

Orientation tolerances are datum related. Perpendicularity, angularity, and parallelism are all orientation tolerances. *Orientation tolerances* may be applied using either MMC or RFS. RFS is implied per Rule #2. Unless otherwise specified, all geometric controls apply RFS. When applied to a surface, orientation tolerances also control form (flatness and/or straightness) tolerances within the orientation tolerance value. They must be contained within the size tolerances. When applied to a feature of size with an axis or centerplane, a virtual condition or inner/outer boundary will occur, which may extend beyond the limits of size. Rule #1 then would not apply.

Carefully consider your design before applying a control *directly* to an axis or centerplane as there may be more than one feature sharing the same centerline or centerplane. It may not be clear which feature the control was intended for (i.e., a counterbore and drilled hole or holes hidden behind the feature in question sharing the axis).

Perpendicularity

Perpendicularity is the condition of a surface, center plane, or axis at a right angle (90°) to a datum plane or axis. A *perpendicularity tolerance* specifies one of the following:

- A tolerance zone defined by two parallel planes perpendicular to a datum plane or axis within which a surface or median plane of the considered feature must lie.
- A tolerance zone defined by two parallel planes perpendicular to a datum axis within which the axis of the considered feature must lie.
- A cylindrical tolerance zone perpendicular to a datum plane within which the axis of the considered feature must lie.
- A tolerance zone defined by two parallel lines perpendicular to a datum plane or axis within which an element of the surface must lie.

Figure 4-2 illustrates surface perpendicularity to only one datum. This allows the controlled surface to be skewed within

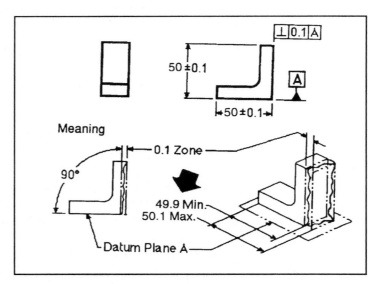

Figure 4-2.   Perpendicularity—surface.

the size limits to other surfaces. This condition is called *degrees of freedom*. With only one datum invoked, we still have full freedom of disorientation to the other two datum surfaces which may require a secondary datum, or other orientation control.

Figure 4-3 illustrates perpendicularity of an axis at both RFS and MMC. Note that with RFS, the tolerance is always the same (.003), while the outer boundary is variable (.503 to

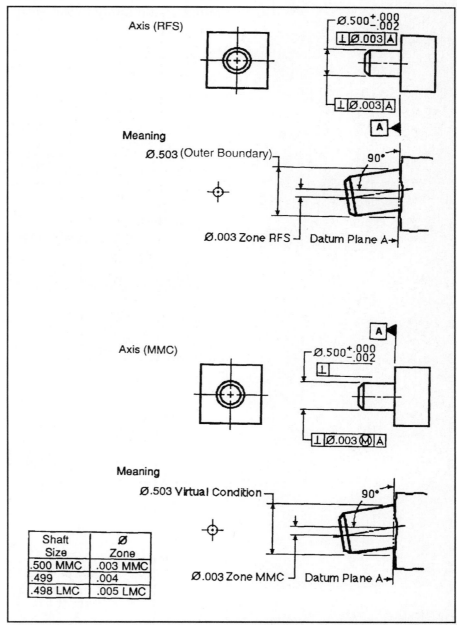

Figure 4-3.   Perpendicularity—axis (RFS) versus axis (MMC) to datum surface.

.501). With MMC, the tolerance is variable depending on size, while the virtual mating size is constant (.503) at the worst case or closest fit. In the MMC example, perpendicularity is applied to a feature of size. Therefore, a virtual condition may occur.

Figure 4-4a shows the use of zero tolerance at MMC and the effect of a restrictive control, which does not allow the full bonus tolerance to be used. Figure 4-4b illustrates the application and control of radial line elements. If the qualifying note (each radial element) were not used beneath the control frame, the control callout would apply to the entire surface.

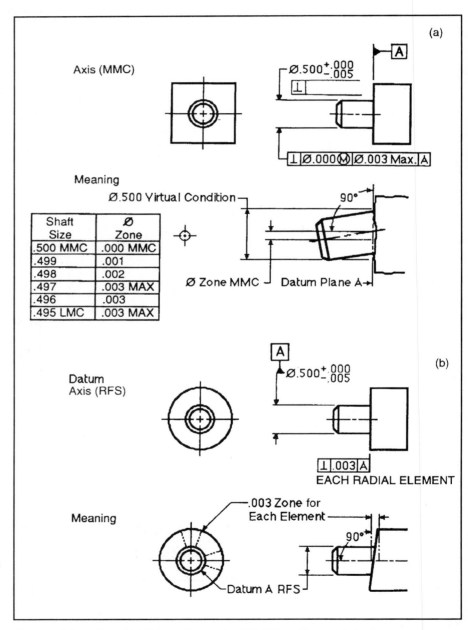

Figure 4-4.   Perpendicularity—axis (MMC) with restricted tolerance—and axis as primary datum.

Angularity

Angularity is the condition of a surface, center plane, or axis at a specified angle (other than 90°) from a datum plane or axis. *Angularity tolerance* specifies one of the following:

- A tolerance zone defined by two parallel planes at the specified basic angle from a datum plane, or axis, within which the surface or center plane of the considered feature must lie.
- A tolerance zone defined by two parallel planes at the specified basic angle from a datum plane, or axis, within which the axis of the considered feature must lie.
- A cylindrical tolerance zone whose axis is at the specified basic angle from a datum plane, or axis, within which the axis of the considered feature must lie.
- A tolerance zone defined by two parallel lines at the specified basic angle from a datum plane, or axis within which a line element of the surface must lie.

Angularity is applied with the same basic principles as perpendicularity, except angularity is intended for use with features related to datums at orientation other than 90°. (See Figures 4-5, 4-6, and 4-7.)

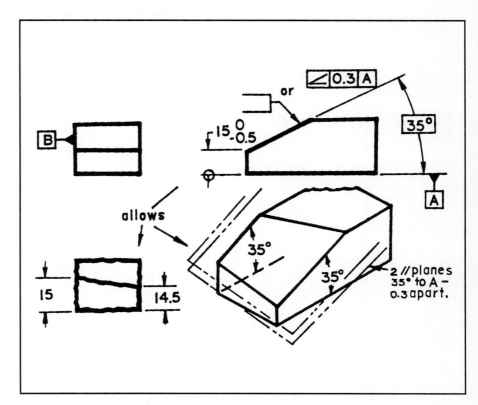

Figure 4-5. Angularity of a surface.

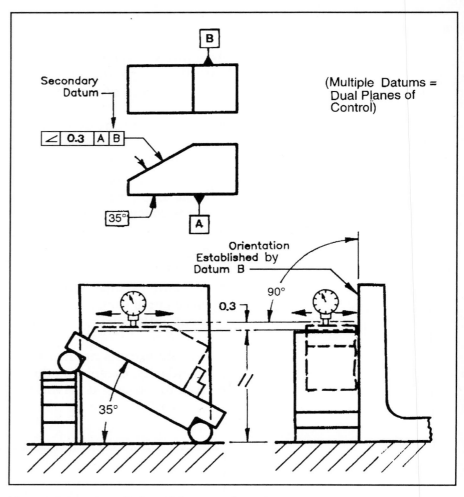

Figure 4-6.   Angularity with secondary datum.

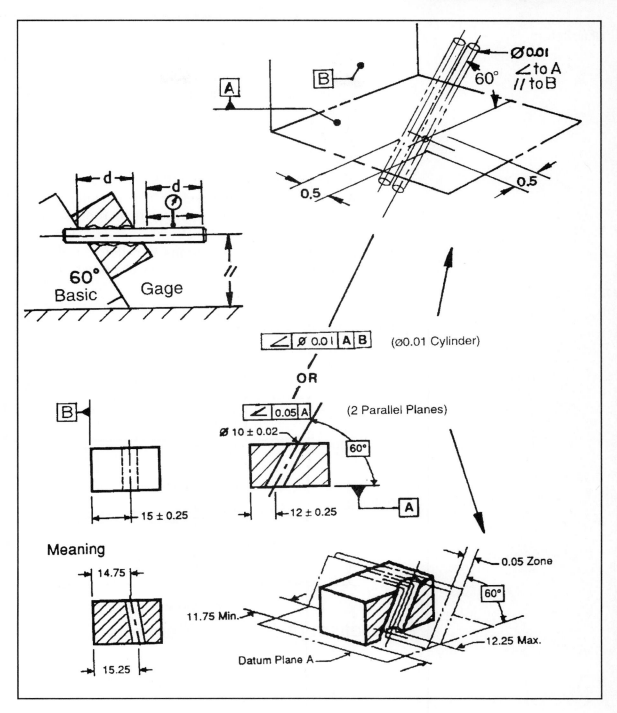

Figure 4-7.   Angularity of an axis.

Parallelism

Parallelism is the condition of a surface or center plane equidistant at all points from a datum plane or an axis equidistant along its length to a datum plane or axis. *Parallelism tolerance* specifies:

- A tolerance zone defined by two parallel planes parallel to a datum plane, or axis within which the surface or center plane of the considered feature must lie.
- A tolerance zone defined by two parallel planes parallel to a datum plane or axis within which the axis of the considered feature must lie.
- A cylindrical tolerance zone parallel to a datum plane or axis within which the axis of the feature must lie.

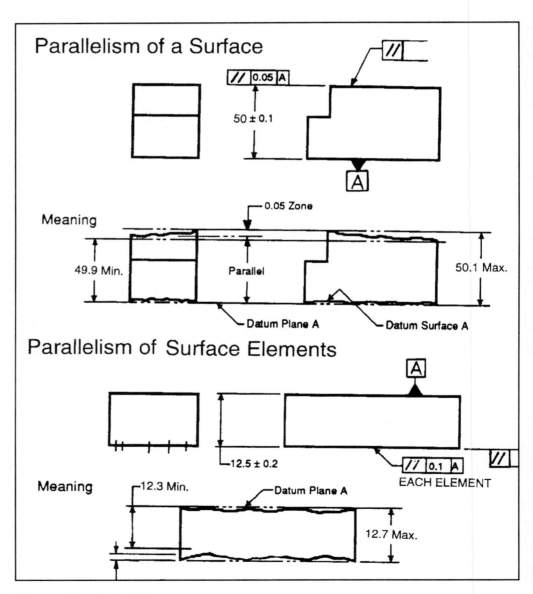

Figure 4-8. Parallelism.

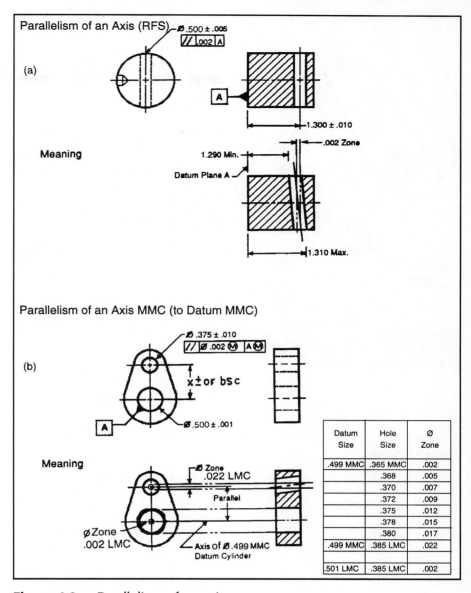

Figure 4-9. Parallelism of an axis.

• A tolerance zone defined by two parallel lines parallel to a datum plane, or axis within which an element of the surface must lie.

Parallelism, perpendicularity, and angularity have similar principles governing their use. The placement of the control frame is optional, as it is with perpendicularity and angularity. When applied to a feature surface, the controlled feature must be within size limits as before. (See Figures 4-8, 4-9, and 4-10.)

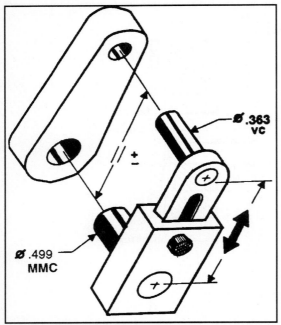

Figure 4-10. Illustration of possible functional gage for Figure 4-9b.

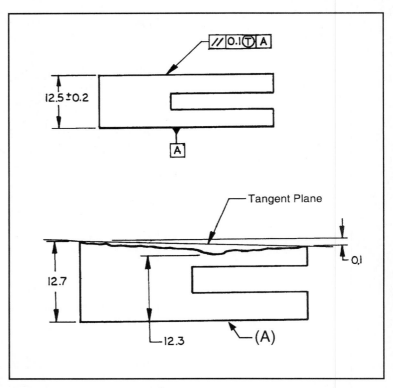

Figure 4-11. Tangent plane.

Tangent Plane

The concept of *tangent planes* has been added to the ASME Y14.SM-1994 standard. It differs from parallelism in that the tangent plane control allows a plane contacting at the high points of a surface to be within the parallelism tolerance control limit, with the surface elements controlled only by size tolerance. Parallelism requires *all* surface elements to be within size, as well as within the parallel limits. This is a subtle design difference and should be reviewed carefully. (See Figure 4-11.)

Where it is desired to control the orientation of a feature surface as established by contacting points of that surface, the tangent plane symbol is added within the feature control frame. A plane contacting the high points of the surface shall lie within 2 parallel planes 0.1 apart parallel to datum A. The surface must be within the specified limits of size.

Carefully consider your design before applying a control directly to an axis or centerplane as there may be more than one feature sharing the same centerline or centerplane. It may not be clear which feature the control was intended for (i.e., a counterbore and drilled hole or holes hidden behind the feature in question sharing the axis.)

EXERCISE 4-1. PARALLELISM

1. Parallelism is applied MMC unless otherwise specified.   T   F
2. Parallelism is datum related.   T   F
3. On a flat surface, parallelism also controls _____.
4. Parallelism should be _____ than size tolerance. (Use less than, in addition to, or greater than.)
5. A parallelism callout specifies a ± tolerance zone.   T   F

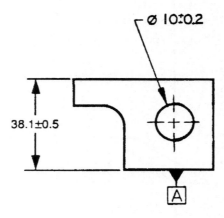

6. On the figure above, indicate a parallelism tolerance of 0.2 to datum A. Draw the tolerance zone (top surface).
7. Indicate the axis of the hole to be parallel to datum A within 0.1 diameter at MMC.
8. What is the size of the hole tolerance zone in the figure above if the diameter is at LMC? _____
9. What is the maximum (total) value the 38.1 dimension can be? _____
10. What is the hole virtual condition *size* limit at MMC? _____
11. When applied to a flat surface, the parallelism tolerance is additive to size tolerance.   T   F

EXERCISE 4-2.   PERPENDICULARITY

1. Perpendicularity is applied RFS unless otherwise specified.
   T   F

2. Perpendicularity is datum related.
   T   F

3. Virtual condition is invoked when perpendicularity is applied to a feature _____ or _____.

4. Perpendicularity applied to a *flat* surface also controls _____ as well as _____ of surface elements.

5. Perpendicularity may be applied to cylindrical feature relationships.
   T   F

6. The figure below requires a 90° angle dimension.
   T   F

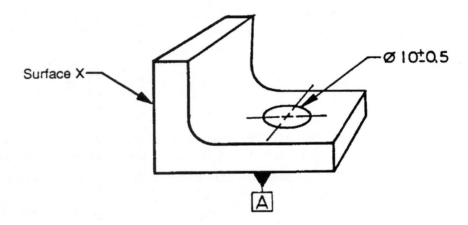

7. On the figure above, indicate surface X to be perpendicular to datum A within 0.1. Draw the tolerance zone.

8. Indicate the hole to be perpendicular to datum A within a diameter of 0.2 at MMC.

9. What is the size of the tolerance zone in the figure above if the feature is 10.5 diameter?

   _____

10. What is the virtual condition size of the hole at MMC?

   _____

EXERCISE 4-3.   ANGULARITY

1. Angularity is applied MMC unless otherwise specified.   T   F
2. Angularity is not datum related.   T   F
3. An angularity tolerance zone is: (a) (2) parallel planes or (b) a wedge shaped zone.
4. Angularity may be applied to a surface at MMC.   T   F
5. Angularity should be used to control the relationship of two 90° surfaces.   T   F
6. To control the hole in the figure below, a feature of size, the use of MMC will allow a bonus tolerance.   T   F
7. For the figure below, indicate the hole to have a 60° relationship to datum A, and provide an angularity callout of 0.1 diameter to datum A. Draw the tolerance zone.

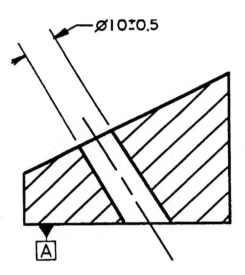

8. The LMC hole size is _____.
9. What is the virtual condition size of the hole at MMC?_____
10. Illustrate that the inclined surface is 25° basic relative to datum surface A, with an angularity tolerance of 0.2 to A.
11. Illustrate the tolerance zones in the figures below.

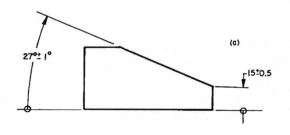

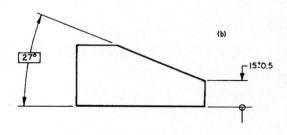

5

Profile Tolerancing

Profile line surface (RFS)

Profile Tolerance is a method used to specify a permissible deviation from the desired profile, usually an irregular shaper where other geometric controls are inappropriate.

Figure 5-1. Profile.

The *profile tolerance* specifies a uniform boundary along the true profile within which the elements of the surface must lie. It is used to control form or combinations of size, form, orientation, and location. Where used as a refinement of size, the profile tolerance must be contained within the size limits.

I consider *profile tolerancing* to be one of the most versatile controls available today. Profile controls may be used:

- with or without datums;
- to control size;
- to control curved or warped surfaces; or
- to control the coplanar relationship of flat surfaces.

In addition, when used as a composite control, profile allows a degree of position and form control within the same callout. Profile controls may be either line or surface controls and are recognized by the symbols shown in Figure 5-1. Profile generally applied RFS to features, but may be applied RFS or MMC to the datums as required. (See Figure 5-6.)

For line profiles, the tolerance zone exists at any section parallel to the view in which it is presented. For surface profiles, the tolerance zone applies to the entire surface in the view presented. The examples in this section show surface profile symbols, however, line profile symbols also could be used.

Normally, the profile tolerance zone is equally disposed about the true basic profile and identified with basic dimensions. But it doesn't have to be. To dispose the tolerance unequally, simply set your dimensions for one side of the tolerance zone and the remaining tolerance will apply to the opposite surface (see Figure 5-2). Also note that this method can be used to apply all the tolerances to either the inside or outside of the true profile (see

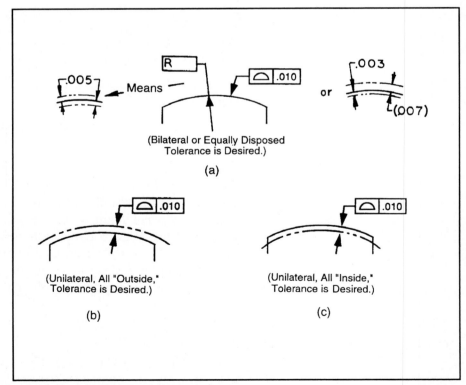

Figure 5-2.   Profile tolerance is a method used to specify a permissible deviation from the desired profile, usually an irregular shape where other geometric controls are inappropriate.

Figures 5-2b and 5-2c). In some designs, this may be helpful in insuring that the tolerance does not inadvertently get added to the maximum size limits (see Figure 5-5). Unless otherwise specified, the interpretation of Figure 5-2a applies to profile controls.

Figure 5-3 illustrates the use of *profile control* in order to also control size. No datums are invoked and the tolerance is equally disposed all around the true basic profile (note the all around symbol). The profile control applies only in the view shown as there is no automatic perpendicularity of surfaces. The control has no datums invoked so the true profile has no orientation to other surfaces.

Figure 5-4 is similar to Figure 5-3. It is modified to illustrate the use of a datum framework. Two profile controls are specified—the curved ramp surface between points X and Y, and the perpendicular surface between points Y and Z. In this illustra-

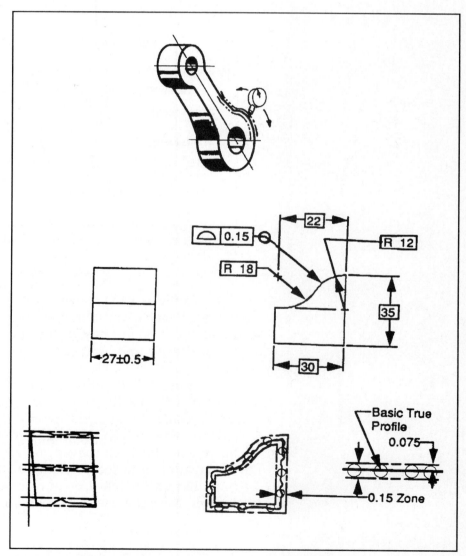

Figure 5-3. Profile of a surface and size control.

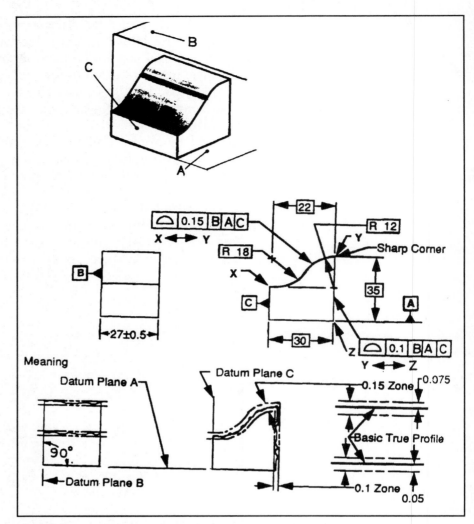

Figure 5-4.   Profile control form and orientation.

tion, profile is used to control form and orientation (perpendicularity). Because datums have been invoked, the profile tolerance zones must be oriented to the primary datum, B, to control perpendicularity and located from secondary and tertiary datums, A and C.

Figure 5-5 illustrates the use of profile as a refinement of size with the profile tolerance zone free to float. Without datums in the control frame, the profile tolerance zone may tilt or twist (within size limits). In addition, the designer has specified the line elements of the curved surface between the points X and Y to be parallel to datum surface A. This type of multiple controls combination for features is common to achieve multi-directional control. Note the use of the chain line to illustrate that all the tol-

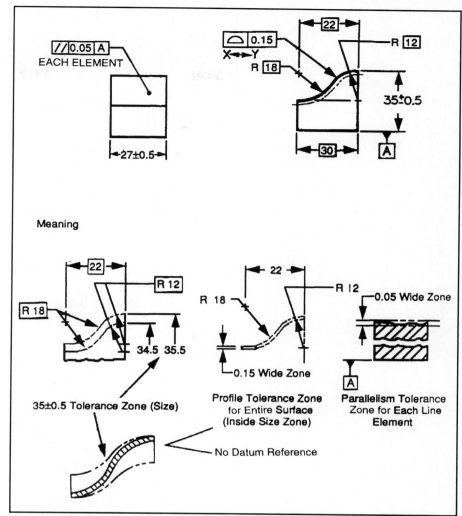

Figure 5-5.   Profile within size limits and combined with other controls.

erance is internal to the basic true profile. In this example, the 35±0.5 not only controls the limits of size, but also locates the start of the true profile R12 radius.

Profile Applied—Secondary and Tertiary Datums
MMC versus RFS

Figure 5-6a illustrates the use of two holes of a pattern of four which are used as the secondary and tertiary datums B and C. The internal and external boundaries are controlled to the datum frame by use of profile control at MMC for the secondary and tertiary datums. In Figure 5-6a, the datum holes are at MMC so a

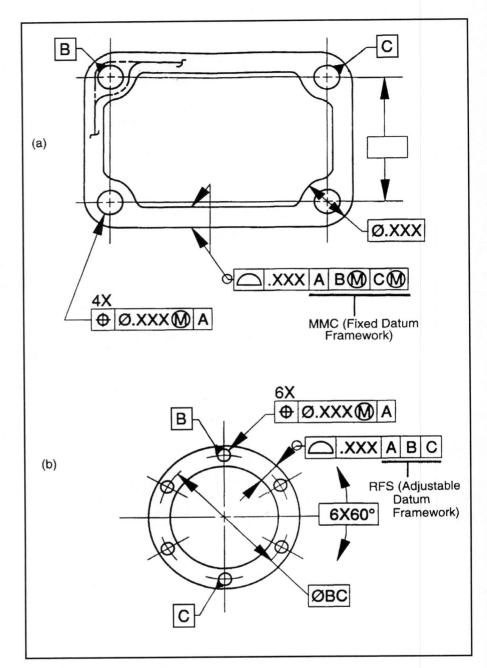

Figure 5-6. Profile—datums MMC and RFS.

functional fixed pin type gage may be used. But in Figure 5-6b, the profile callout is RFS, thus requiring the gage pins and fixtures to be adjustable regardless of the size of B and C. We saw some of the effects of MMC and RFS on datums in the datum section. Figure 5-6 may be typical of some gasket and shim stock dimensioning.

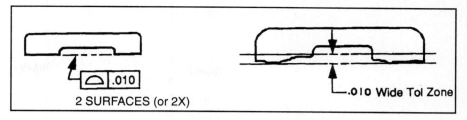

Figure 5-7.   Profile for control of coplanar surfaces.

As noted previously, profile may be used to control coplanarity of two or more flat surfaces. Coplanarity is the condition of two or more surfaces having all elements in one plane. Profile tolerance for coplanar surfaces is a profile of a surface tolerance which may be used where the designer desires to treat two or more surfaces as a single interrupted or noncontinuous surface. In this case, a control is provided similar to that achieved by a flatness tolerance applied to a single plane surface. The profile of a surface tolerance establishes a tolerance zone defined by two parallel planes within which the considered surfaces must lie. A common technique on older drawings was the use of a note such as "Surfaces A and B to lie in the same plane within .010." The use of profile control as shown in Figure 5-7 simplifies and symbolizes this requirement. Remember flatness applies to a single feature (surface).

Profile may also be used on multiple, coplanar surfaces. Figure 5-8 illustrates the use of any two coplanar feet as datums

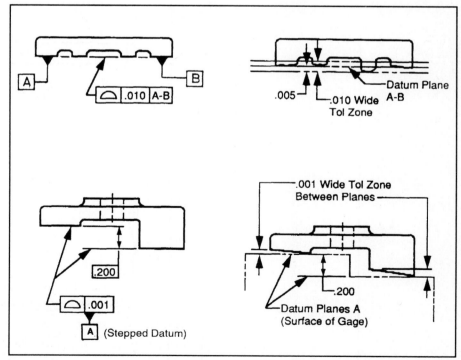

Figure 5-8.   Profile control (coplanar and stepped surfaces).

A and B with the remaining feet controlled relative to these datums. Profile may be used to control and identify datums of two parallel, offset feet as well. This is a stepped datum condition. The surface control of .001 applies to both surfaces equally, as they are considered as one. Placement of the datum symbol confirms this.

Composite Profile Tolerancing

In *composite profile tolerancing*, the upper callout is the profile boundary location control. It governs the location of the profile boundary. The lower callout is the size, form orientation refinement and specifies the smaller boundary for the feature which must be maintained within the location limit of the upper callout. The actual feature surface must fall within both zones.

Composite profile tolerancing is much like composite position tolerancing, except that with profile, we are dealing with a bound-

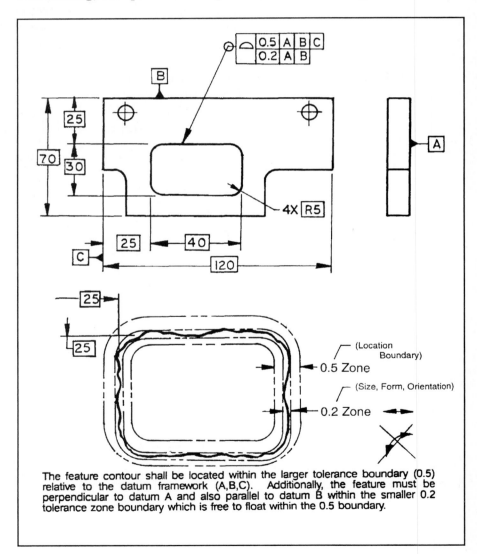

The feature contour shall be located within the larger tolerance boundary (0.5) relative to the datum framework (A,B,C). Additionally, the feature must be perpendicular to datum A and also parallel to datum B within the smaller 0.2 tolerance zone boundary which is free to float within the 0.5 boundary.

Figure 5-9.   Composite profile control.

ary tolerance zone rather than a centerplane or axis tolerance zone (see Figure 5-9). In Figure 5-9, the smaller tolerance is relative to the A and B datums, which means the C datum is the only control removed in the lower callout. Therefore, the 0.2 zone can slide horizontally, but must remain perpendicular to A and parallel to B. The tolerance zone cannot twist or rotate. The tolerance zone must remain within the 0.5 zone.

Multiple Profile Controls

Adapting the datum and profile control principles established earlier, it's possible to use profile for various design controls (see Figure 5-10). This figure illustrates the use of profile controls on complex shapes created by mathmatically defined datums. Composite profile controls are used and extended to include both datum references and no datum references (form control). The top control establishes the datum framework and, therefore, controls location. The second control callout is relative to datums A and B, and dictates a 0.4 tolerance zone perpendicular to A and

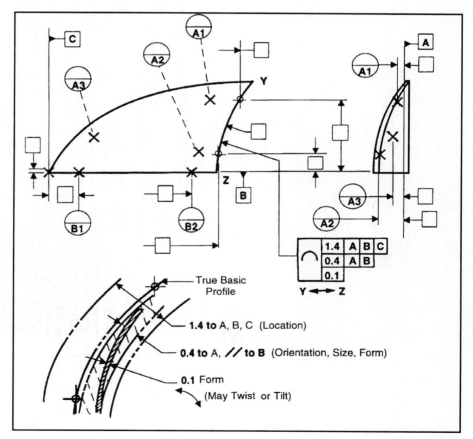

Figure 5-10.   Multiple controls complex surface (element).

parallel to B. This may slide parallel to B within the limits established by the 1.4 location tolerance zone. The lower callout is a further refinement of form without regard to any datum reference. Therefore, the 0.1 zone may tilt or twist and float within the limits established by the 0.4 zone. The 0.1 tolerance zone has all degrees of freedom.

The Boundary Principle

Profile tolerancing may be used in combination with position tolerancing when you desire to control the location of irregular shapes to a datum framework. Position tolerancing is used with features subject to size variation and which normally have a centerline or centerplane (holes/slots).

If the feature is irregular in shape and does not fit the definition of *feature of size* (cylinder, sphere, or two paraplane surface),

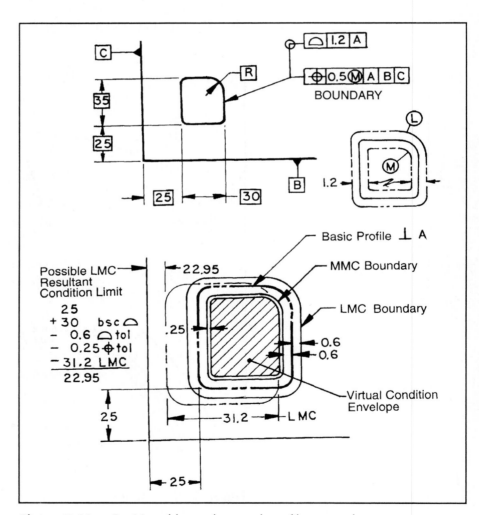

Figure 5-11.   Positional boundary and profile control.

you may want to consider the use of profile tolerancing to control size and shape along with position tolerancing for location control. This concept is illustrated in Figure 5-11.

The profile tolerance controls the size and shape, and is not related to any datum framework other than perpendicularity to datum A. The position control locates the shape, at the basic dimensions shown, when the feature is at its basic size.

The MMC principle allows a bonus location tolerance when the feature departs from MMC and gets larger. As the feature reaches LMC, the resultant condition creates a minimum wall thickness of 22.95 as shown in Figure 5-11.

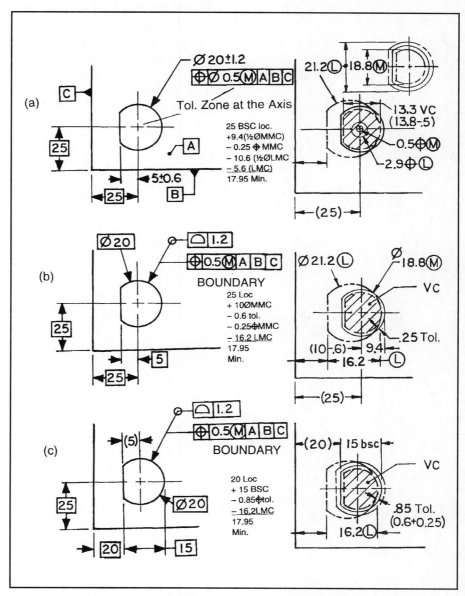

Figure 5-12.   Positional boundary dimensioning techniques.

Figure 5-12 offers further explanation of the *boundary principle* when used with other combinations of plus-minus profile and position controls to both feature centerlines, centerplanes, or surfaces.

Figure 5-12a illustrates plus-minus size dimensioning along with positional tolerance controls to a feature axis (centerline). This dimensioning and tolerancing combination will yield a MMC diameter of 18.8 and a LMC diameter of 21.2. The flattened surface will be 5.6 maximum and the positional limit at LMC will be 2.9 diameter with the resulting minimum wall thickness of 17.95.

Figure 5-12b illustrates a basic 20 diameter and profile tolerance for size and shape, plus positional tolerancing with the boundary control for location of feature surface. As shown, these combined controls allow the same MMC and LMC boundaries, as well as the same minimum wall as Figure 5-12a.

Figure 5-12c illustrates a basic 20 diameter and profile tolerance. The position boundary limit is established by the feature surface. With this dimensioning and tolerancing technique, the MMC and LMC limits are the same as shown in Figures 5-12a and 5-12b. By using a surface as the basic profile positional limit, the resultant minimum wall thickness is also 17.95.

EXERCISE 5-1. PROFILE (LINE/SURFACE)

1. Profile is applied RFS unless otherwise specified.   T   F

2. Profile tolerancing is always datum related.   T   F

3. Profile also may be used to control the _____ of cylindrical surfaces or _____ relation of flat surfaces.

4. ⌒ is the symbol for surface (depth) control.   T   F

5. Profile may be used in combination with other geometric controls or with size controls.   T   F

6. In certain applications, the profile tolerance zone may extend beyond size limits.   T   F

7. A datum or datums may be used to orient the tolerance zone to a surface or surfaces.   T   F

8. The profile tolerance zone is understood to be unilateral.   T   F

9. Indicate the surface (between points X and Y) in the figure below to be relative to datum A primary and datum B secondary within 0.2. Draw the tolerance zone.

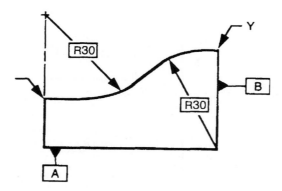

10. Indicate the true profile of the surface in the figure above to be within 0.1 between X and Y.

EXERCISE 5-2.   SPECIFYING PROFILE OF A SURFACE BETWEEN POINTS

1. Complete the drawing control callouts ① and ② to satisfy the requirements indicated.

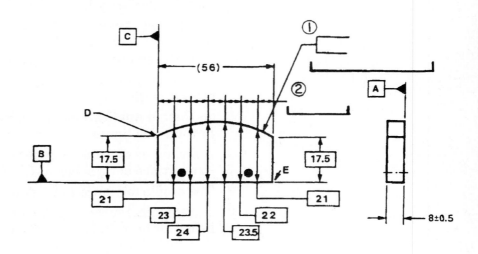

The surface between points D and E must lie between two profile boundaries 0.25 apart, perpendicular to datum plane A, equally disposed about the true profile and positioned with respect to datum planes B and C.

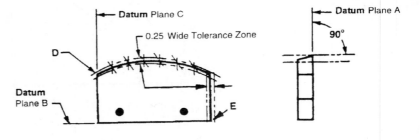

6

Runout

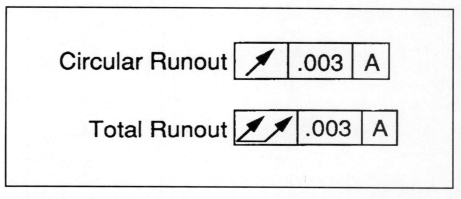

Figure 6-1.   Runout.

Runout is a composite control used to specify the functional relationship of one or more features of a part to a datum axis. *Runout tolerance* indicates the permissible error of the controlled feature surface when rotated about a datum axis. The specified tolerance of the controlled feature indicates the maximum full indicator movement (FIM) when the part is rotated 360°.

Runout is a composite feature control which is applied to a feature surface RFS. It can be applied to a single circular element or to the entire surface. *Circular runout* is commonly understood, while *total runout* is generally described as containing enough readings to give reasonable assurance that the surface has met the requirements specified. With total runout, the indicator device must have the ability to traverse the surface, perpendicular or parallel to the primary datum. The total runout evaluation method is considered a quality engineering decision, based on capability, past performance data, risk and established quality standards.

Circular runout is a single element reading as illustrated by Figure 6-2a. The primary datum is normally an axis. All mea-

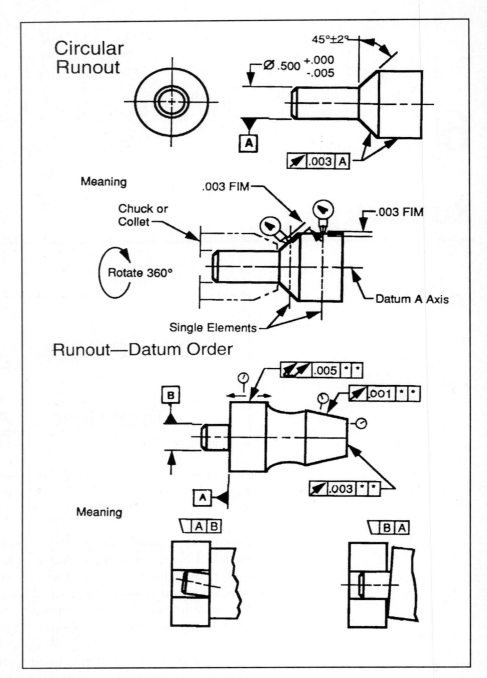

Figure 6-2.   Circular runout and datum order.

surements are taken normal to the surface being measured as shown. *Any* circular reading may be used, as determined by the quality engineer or as specified by the drawing. Figure 6-2b shows the potential effects of reversing (intentionally or unintentionally) the datum order. The measurement results can be totally different on the same part.

Figure 6-3a illustrates the requirement of total runout and the necessity of moving the indicator to evaluate the entire surface. All measurements are taken normal to the measured surface as before. Figure 6-3b further illustrates the need to evaluate datum features to the common axis of a shaft. Opposite ends of the shaft have been designated datums A and B, but, in addition,

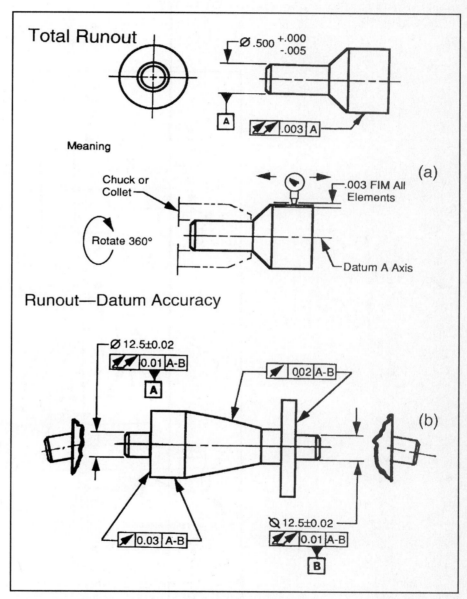

Figure 6-3.   Total runout and datum accuracy.

the datums themselves must also be evaluated, in order to properly control all the elements in the shaft design. Figures 6-4 and 6-5 illustrate possible set-ups for evaluation, along with datum terms which we have discussed previously.

In Figure 6-4, note that the shaft ends (tool centers) were used as primary datums A-B. In this example, the secondary datum is automatically established, but not identified, as no lateral movement (float) can occur. The part is locked in the measurement fixture, therefore, all readings are direct or actual values.

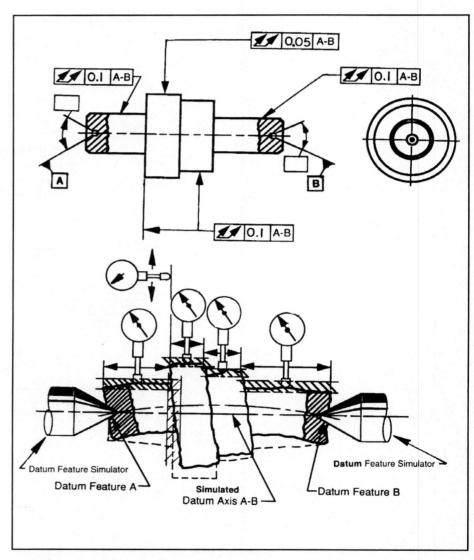

Figure 6-4.   Datums—matching centers.

Now look at Figure 6-5. The datums are axis A-B primary, with surface C secondary. This set-up allows lateral movement which could influence the design integrity as well as measurement repeatability and part acceptance. Looking further at Figure 6-6, we see the problem of using a shoulder as a secondary datum or stop. If secondary datum feature C is not perpendicular to the primary datum feature axis, lateral float could occur and be read in the FIM reading of surface A, as well as other vertical surfaces (see Figure 6-6). This condition is exaggerated and shown in Figure 6-6b and c, to illustrate the impact of the location of the stop on surface C.

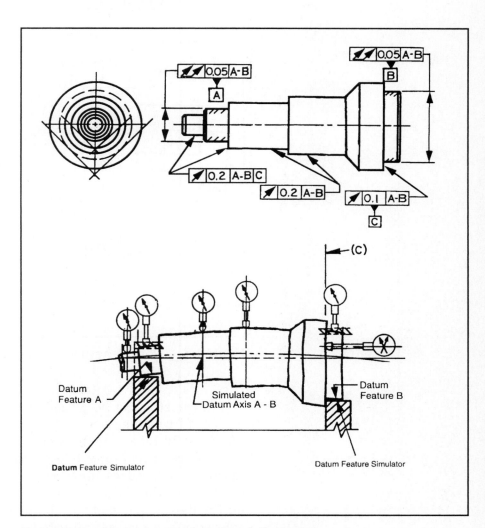

Figure 6-5.   Datums—two functional diameters.

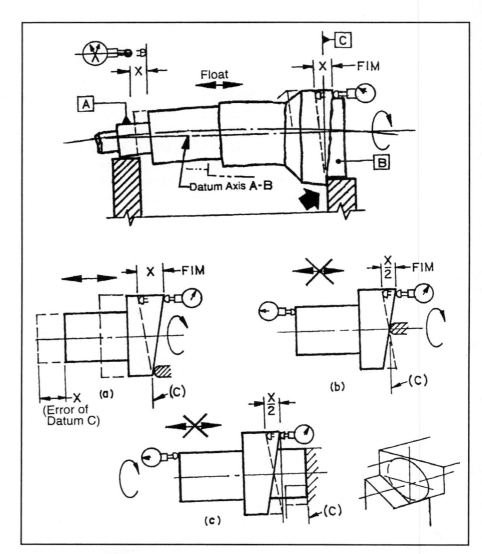

Figure 6-6.   Possible results.

If the stop can be placed at the axis as shown in Figure 6-6b and c, all readings are actual, as no float occurs or impacts the readings. If, however, the stop is located at the edge of the shoulder, the readings may potentially double due to the float. The "good" part features may be ruled as out of conformance due to the float of the datum. When this condition exists, the readings for the datum generally are halved. The datum error must be considered when evaluating other features relative to their conformance. Secondary datum error can not be avoided and, therefore, must be minimized or otherwise allowed for when evaluating other features.

Datum Errors/Vee Locators and Runout

When evaluating runout, care must be taken to first find or resolve any *error which may exist in the datum.* Datums can have size, form, and position error. Therefore, if vee locators are used for the set-up, we may have a perfectly concentric feature, read the error produced by the ovality of the datum, and possibly reject the part for the wrong reasons. Further, the location of the indicators can give us varied results. In addition, the included angle of the vee locators may impact the measurement results as shown in Figures 6-7 and 6-8.

We might read very little vertical error and considerable horizontal error with 90° vee blocks. With 120° vee blocks, the vertical and horizontal errors tend to even out. These illustrations point out the importance of knowing and understanding the set-up method, indicator location, vee locator included angle, and datum size and shape when evaluating measurement results relative to runout when using vee locators for the datum simulators.

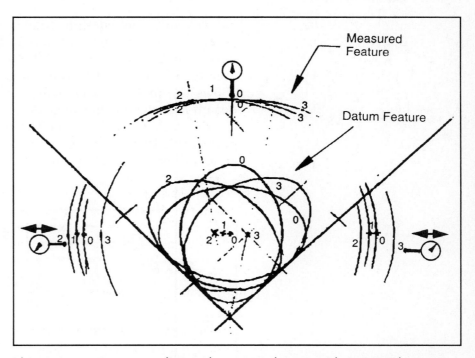

Figure 6-7. Datum ovality and measured error with 90° vee locators.

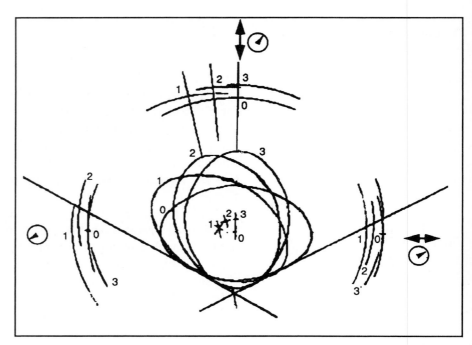

Figure 6-8.   Datum ovality and measured error with 120° vee locators.

Primary Datum Surface

Occasionally, some designs may require features to be controlled to primary datum surfaces using the diameter as the secondary datum as shown by Figure 6-9. The figure illustrates a collar diameter to be controlled to the datum framework C, D. This callout shows the primary design influence as surface C with secondary datum D perpendicular to C. Secondary datum D is used for centering the part in order to take the runout measurements. Secondary datum D is RFS. Remember the datum rule: Any out of squareness error will be reflected in the secondary and/or tertiary datums.

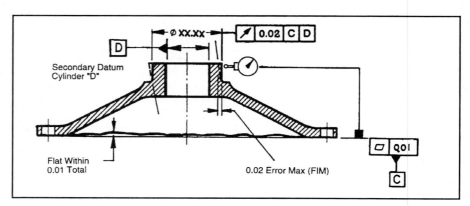

Figure 6-9.   Part mounted on large flat surface (datum) and diameter (datum).

EXERCISE 6-1.   RUNOUT (CIRCULAR/TOTAL)

1. Runout is understood to apply at MMC.   T   F
2. Circular and total runout callouts control _____ elements and include the errors of _____ and _____.
3. Additionally, total runout controls _____ of a cyclindrical surface.
4. Runout may be used with or without datums.   T   F
5. Runout controls are read at a feature surface and are relative to a datum axis.   T   F
6. The control of a convex or concave *surface* to a datum axis can be accomplished with circular runout.   T   F.

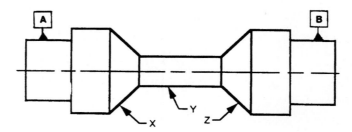

For the figure above:

7. Indicate a total runout to datum A of 0.2 for surface X.
8. Indicate surface Z to have a circular runout to datum B of 0.1.
9. Indicate surface Y to have a total runout to datum A and B of 0.05.
10. Draw the tolerance zone(s).
11. Runout may be applied to a surface, relative to a primary flat datum surface and secondary datum axis.   T   F

7

Location Tolerances

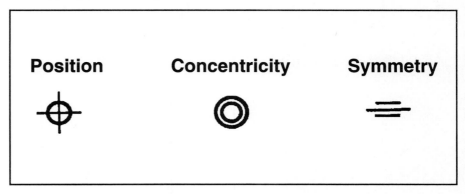

Figure 7-1. Location tolerances.

Location Tolerance Controls

There are three location tolerance controls—*position*, *concentricity*, and *symmetry*. These controls all deal with the control of features that normally have a centerline or centerplane. Position may be applied MMC, LMC, or RFS, while concentricity and symmetry are applied only RFS. RFS is always implied unless otherwise stated. We will begin with position as it is very versatile. We'll then move on to cover concentricity and symmetry.

True position is a term used to describe the perfect (exact) location of a point, line, or plane (normally the center) of a feature in relationship with a datum reference or other feature. (See Figure 7-2.) A position tolerance is the total permissible variation in the location of a feature about its true position. For cylincrical features (holes and bosses) the position tolerance is the *diameter (cylinder)* of the tolerance zone within which the axis of the feature must lie. For other features (slots, tabs, etc.) the position tolerance is the *total width* of the tolerance zone within which the center plane of the feature must lie.

A feature (hole) is rarely produced to a perfect location. To repeat an exact hole location is even more difficult. Figure 7-2a illustrates possible locations of a single hole produced repeatedly in a loose fixture.

Most parts contain patterns of holes, located as a pattern, with each hole having a location tolerance as shown in Figure 7-2b.

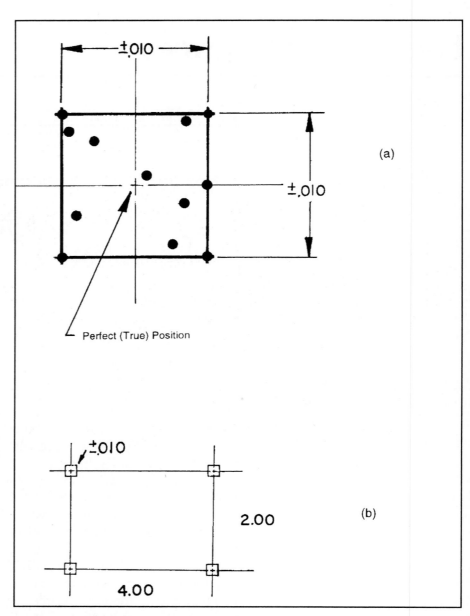

Figure 7-2. Perfect (true) position.

Each axis represents a true center. Each square represents the ±.010 tolerance zone for each hole.

In addition to the hole (individual) tolerance, the pattern of four holes must be located in a part, relative to some functional feature(s) such as edges or centerplanes. Figure 7-3a illustrates a ±.015 tolerance zone for the pattern of holes from the part edges.

Figure 7-3b illustrates the potential combined effects of one interpretation of hole-to-hole tolerance and pattern location tolerance using the coordinate dimensioning system. We'll use this example of a part with a pattern of four holes for a design conversion exercise (see Figure 7-6). But first, let's pursue the position tolerance theory a little further.

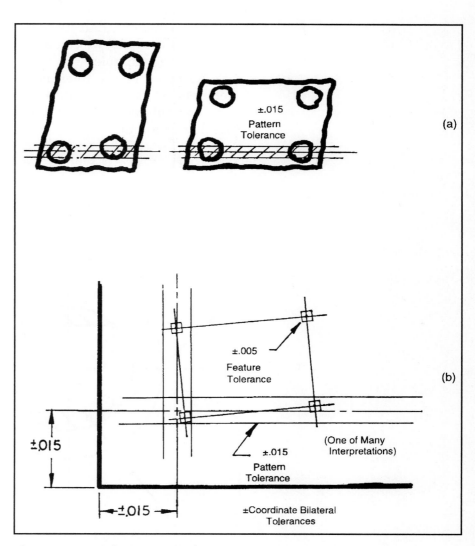

Figure 7-3. Hole patterns tolerances.

Position Tolerance Theory

The position tolerance theory is not that complex as illustrated by Figure 7-4. By converting a common tolerance such as ±.010 (which you might find on a conventionally toleranced drawing) to *positional tolerance*, we find we have gained 57% tolerance area for manufacturing with no sacrifice in quality level. The shaded area was not available with the ± system, however, features (holes) produced at the corners of a ± tolerance zone are actually .014 from the center of perfect location. This converts to a tolerance zone diameter of .028 (1.414 × .020 = .028). Further, with positional tolerancing, we have the possible MMC bonus tolerance, which is size dependent. We did not have this with the conventional system.

The bonus tolerance exists when the feature hole(s) is larger (or shaft smaller) than its MMC size. For every unit increase in the size of the hole, we can add that same amount to the positional tolerance available.

As noted in the tolerancing section earlier, there are three elements which effect mating part fits—hole size, fastener size, and total tolerance. (See Figure 7-5.) Mathmatically, these work together as a unit:

$$T = H - F \qquad H = F + T \qquad F = H - T$$

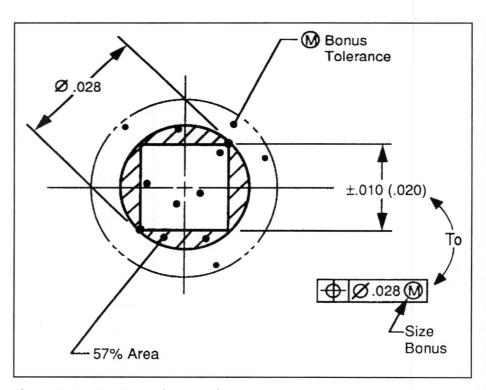

Figure 7-4.   Position tolerance theory.

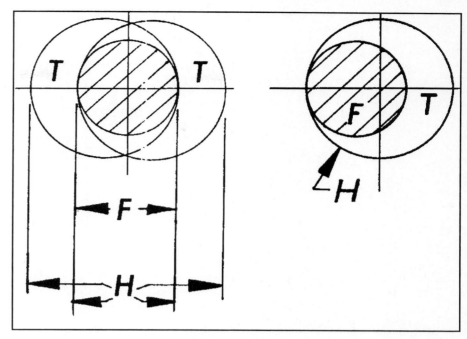

Figure 7-5. Tolerance formula elements.

If one element changes the other elements are effected. Further, the squareness of the features (holes) must be accounted for in calculating the tolerance. This is accomplished by the tolerance zone rule "tolerance zones exist for the full length, width, or depth of the feature."

Position Tolerance Formulas

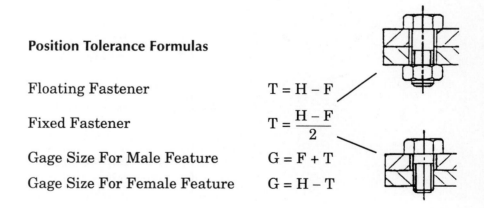

Floating Fastener	$T = H - F$
Fixed Fastener	$T = \dfrac{H - F}{2}$
Gage Size For Male Feature	$G = F + T$
Gage Size For Female Feature	$G = H - T$

Figures 7-6a illustrate a simple four-hole pattern located through the use of the standard rectangular/coordinate dimensioning system for hole locations. We will use this figure as an example and convert the dimensioning/tolerancing to GDT position tolerancing.

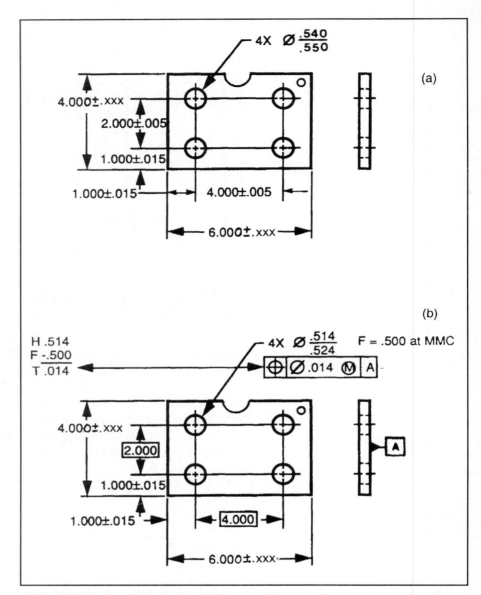

Figure 7-6.   Position—rectangular coordinate tolerance conversion.

First, remember we are not redesigning the part, only converting the tolerancing. In Figure 7-6b, the first step is to convert the hole size and location tolerance of the hole-to-hole tolerance (±.005) to position tolerance of .014 diameter (.010 square × 1.414). The fastener is a .500 diameter bolt MMC, if we add the .014 tolerance to the MMC bolt diameter of .500, we have the minimum clearance hole diameter of .514 diameter. The feature control frame would read: true position of the four holes is .014 diameter at MMC, relative to primary datum A. Note: This is a *floating fas-*

tener design in that both parts have clearance holes. The feature control frame and tolerance controls are shown in Figure 7-6b.

The second step in the conversion concerns the datum framework and the pattern location from that framework. The pattern of four holes is located from the surface, the left and lower edges by the two 1.000±.015 dimensions. Therefore, the edges are the secondary and tertiary datums we must use to locate the hole pattern. Converting the .015 tolerance, we have a diameter tolerance zone of .042, relative to the datum frame A, B, C. This control frame is placed above the previous frame, giving us *composite position tolerance*. The position symbol is entered once and applied to both upper and lower entries. Figure 7-7a illustrates this last step.

The third step in conversion (see Figure 7-7b) could be consideration of datum target points with the possible use of zero position tolerance. Datum targets, as we've discussed, would define the framework precisely. To adjust to zero position tolerance, we make the clearance hole the same as the fastener diameter at MMC (.500). The maximum hole size could be a standard oversize such as .562, rounded to .560. This would allow manufacturing the flexibility of choosing the tolerance by drill or punch selection (i.e., .531 drill = a .031 diameter tolerance zone). With this conversion completed, we've used all available tolerance and precisely defined the datums—requirements much asked for by manufacturing and/or suppliers. A comparison of the tolerance zones is shown by Figure 7-8.

Maximum Hole Size

The maximum size (.560) was pulled out of the air as a reasonable standard oversize. If we take another approach, we may come up with a different number. Consider screw thread designs. For class 2 standard type applications, we can expect threads to be produced to an effective thread form of 55 to 65% (70% considered perfect in ferrous metals). Also, screw heads are considered to have a load factor of two times thread strength. If we use industry standards or *Machinery Handbook* data, we find the bearing area under the screw head to be 15.6mm for a 10mm capscrew (MMC). The difference between these numbers is 5.6mm (or the total bearing area). Half of 5.6 is 2.8 × 70% = 2.0mm oversize hole or 12.0mm for a 10mm bolt. (I use 70% as a comfort margin.)

Thus, a formula could be 1.2 times the fastener diameter at MMC. This formula has worked successfully for me in ferrous metals, but would need adjustment for softer metals, plastics, or other less dense materials. We know that no matter what the for-

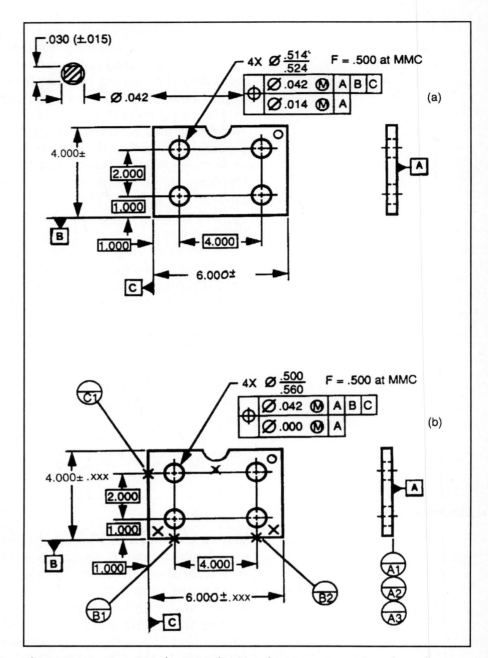

Figure 7-7.   Rectangular coordinate tolerance conversion *(cont.).*

mula would not be less than 1.0. Whatever formula is chosen, we should avoid arbitrary selection of tolerance numbers based only on convenience or past practice. We should be able to justify with some engineering logic the specifications used (see Figure 7-9).

With this information, Figure 7-7b could have a maximum hole diameter of .600, providing that the hole size with appropriate bonus tolerance for position (resultant condition) creates no design problems due to inadequate wall thickness from the part

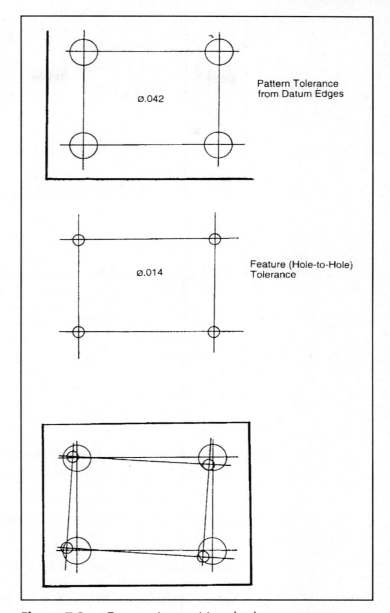

Figure 7-8.   Composite positional tolerance zones.

edges. Always consider the effects of LMC before finalizing design tolerances and control callouts. Holes may be produced with location error, squareness error, or combinations of both. Figure 7-10 illustrates these errors applied to three holes. These errors may be repeated in any number of holes in a pattern of holes.

Tooling/Fixturing

Using the four-hole pattern of Figure 7-7, along with our background of datum targets, we can now apply dimensioning/tolerancing principles correctly so as to insure that design integrity is

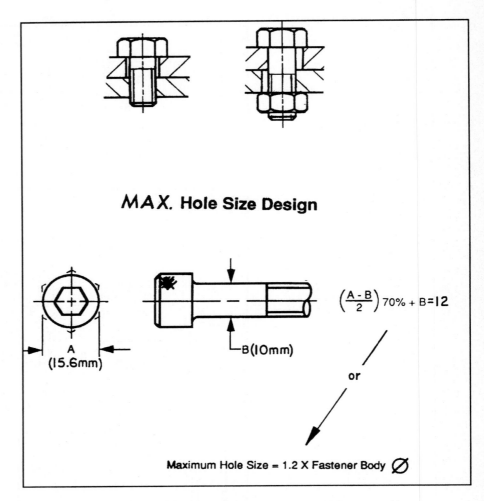

MAX. Hole Size Design

$$\left(\frac{A - B}{2}\right) 70\% + B = 12$$

A
(15.6mm)

B(10mm)

or

Maximum Hole Size = 1.2 X Fastener Body ⌀

Note: This formula typical for general UNC class 2 threads (ferrous metals).

Figure 7-9.   Maximum hole size design.

maintained throughout the manufacturing process. We must consider the impact of design tolerances on manufacturing, tooling/fixturing, and quality control. Design dimensioning and tolerancing principles must be followed to insure interchangeability, measurement repeatability, and compatibility of service replacement parts. Proper targeting/fixturing along with the use of proper tooling/gaging tolerances will also help.

Figure 7-7b illustrates a datum framework and associated target points/lines. These targets may be used by manufacturing and inspection within the limits of tolerance distribution to help achieve product integrity. The processing fixture for the part in Figure 7-7 could look like that shown in Figure 7-11. Most tools used to generate the holes in this particular part move in X and Y directions, thus developing a square tolerance zone not a circular zone as shown by the positional control on the drawing. The drawing callout is a Ø.042 Ⓜ relative to datums A, B, and C.

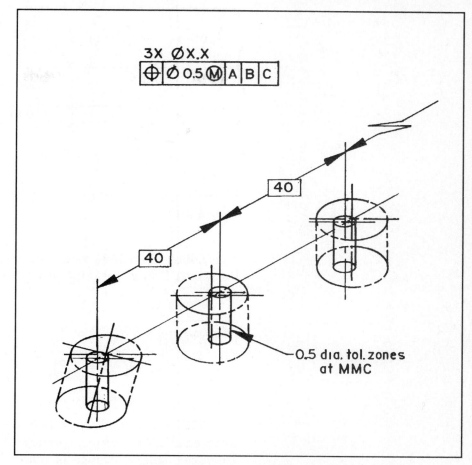

Figure 7-10.   Hole location/perpendicularity error illustrated.

The tool engineer must convert the drawing tolerance to the appropriate tolerance as shown in Figure 7-12. This conversion is generally made RFS.

Going further, the lower control callout must be contained within the above limits. The lower control is .000 Ⓜ, that is, it is

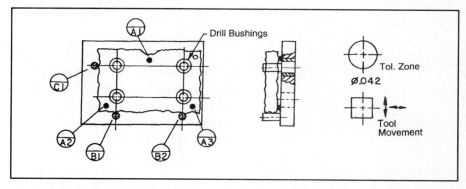

Figure 7-11.   Tooling fixture for Figure 7-7.

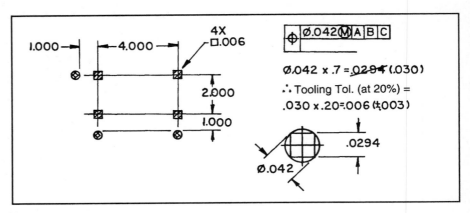

Figure 7-12.

dependent upon hole size. Let's choose a hole size of Ø.530 (.030 oversize). Using the same process as above:

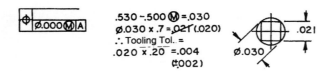

Therefore, the hole-to-hole tolerance of the lower control call-out requires tooling accuracy of ±.002 between drillspindles. However, this tolerance is not related to the fixture targets B1, B2, and C1. The tolerance zones must be perpendicular to fixture surface A.

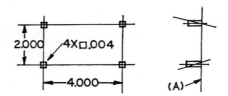

The gage designer or inspector may go through a similar process, except the inspection tolerance will be 5% of the feature size tolerance (ANSI B4.4). Using this example, you'll see that two gages are required—one relative to the datum/targeting framework, the other only perpendicular to datum surface A. Let's follow the process*:

*The process will accept parts with an assembled line fit condition with all variables at the worst case.

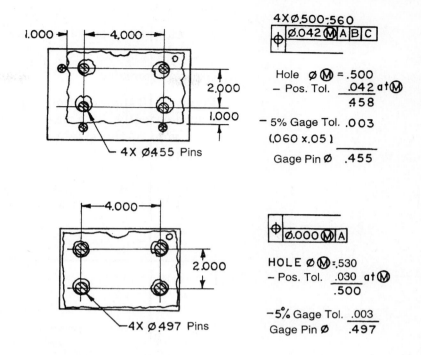

ANSI B4.4 allows the gage design a total of 10% of the feature size tolerance as noted earlier. Generally, this tolerance allowance is divided as 5% for gage design and 5% for gage wear allowance. Further, the allowance is normally divided equally between form and location of the gage features. We need to remember that gages are not always developed (or practical) for the evaluation of parts. Coordinate measuring machines, computer-assisted devices, and surface plate inspection techniques are often used. The total 10% allowance is available to these alternative measurement methods as well.

This exercise is intended to illustrate the need for all areas involved in the product development/manufacturing process to be working to the same set of ground rules. If this, or a similar process, along with existing national standards is used and if tolerances and processes defined in ASME Y14.5M, ANSI B4.4, ANSI B4.1 and 4.2 are followed, we have gone a long way in helping insure design integrity, product interchangeability, and repeatability of measurement results. Service parts compliance will follow.

Position Control Applications

This portion of the text deals with various applications of positional tolerance controls and the possible results. The illustrations shown are not the only possibilities for a particular type control, but will serve as design factors for us to consider.

Feature Symmetry

Figure 7-13 shows optional methods for locating a pattern of holes symmetrically, in a part. By placing the datum symbol as an extension of the size dimensions (Figure 7-13a) we specify the centerplanes of these dimensions as the datums, with the features and the datums at MMC. A RFS control also could have been used. The difference between the callout applying at MMC and RFS is that with RFS, we have no bonus tolerance. Also, the

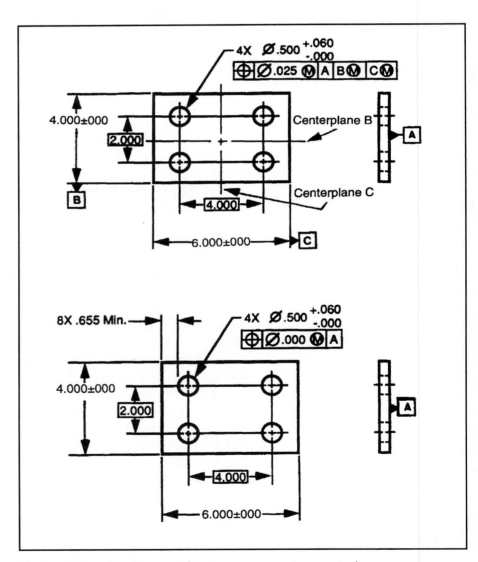

Figure 7-13. Position applications—symmetry control.

gaging must be variable, adjusting to the size of the datums (and features). With MMC, the tolerance is variable with possible bonus. The gaging could be a fixed receiver gage with gage pins. Figure 7-13b illustrates optional dimensioning for a hole pattern and the potential for use of a hole pattern gage (one datum) with a .655 minimum snap gage for insuring symmetry. Other methods are possible.

Multiple Patterns of Holes

Figure 7-14 shows two sets of holes with the same datums, in the same order, and with the same material condition symbols. This technique implies the two sets of holes may be considered as a single pattern for the purpose of design and gaging. If this implication is not the design intent, the notation "SEPT REQT" (seperate requirement) should be added beneath one or both feature control frames as shown in Figure 7- 15.

Multiple Datums Controls

At times, we may have a need to locate features from multiple datum frameworks. The threaded holes of Figure 7-16 are to be controlled from the datum A, B, C framework, in addition, they are to be controlled from the A, D framework. With multiple requirements, we have created more restrictions and limited acceptable combined tolerance zones. To help sort out the toler-

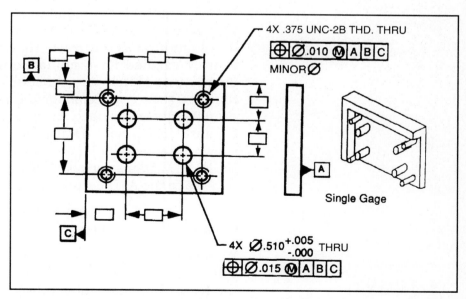

Figure 7-14. Multiple pattern—standard requirement.

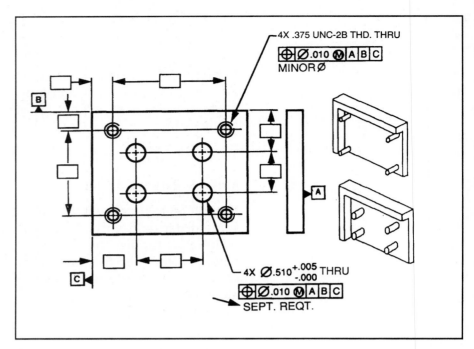

Figure 7-15. Multiple pattern—separate requirement.

ancing and gaging implications of Figure 7-16. Let's review Figure 7-17.

Figure 7-17a illustrates the tolerance zones and a possible functional gage for the two datum D holes and the four threaded .500 diameter holes, both of which are relative to the datum framework A, B, C. As both sets of holes use the same A, B, C framework in one control, these requirements could be expressed in a common gage (see Figure 7-14 for multiple feature requirements).

Figure 7-17b illustrates the two datum holes D could be separately verified for compliance to the position control of .002 diameter at MMC, perpendicular to datum A. Figure 7-17b also illustrates the requirements of the four threaded holes relative to primary datum surface A and secondary datum holes D (RFS). Because datum holes D are expressed in RFS, the gage pins are shown as adjustable and variable. As there is no relationship to the A, B, C framework, the tolerance zones for the four holes are relative to the A, D datum framework.

Conical Tolerance Zones

With very long products, such as shafts or spacers, it may be difficult to hold the position tolerance for the full length or depth of the feature (fundamental rule-tolerance zones). Due to drill run from inconsistent material composition, or for other reasons, we

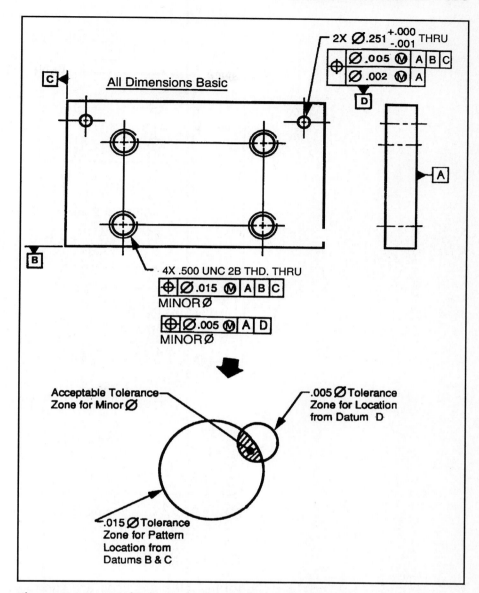

Figure 7-16.   Multiple controls—single features.

may need to allow a larger tolerance at the drill hole exit than was allowed at entry.

The controls applied in Figure 7-18 are one solution if this condition should occur. This control callout provides a 0.5 diameter tolerance zone at entry (surface B), with a 1.0 diameter tolerance zone at surface C. Care must be taken that this technique does not allow insufficient wall thickness at surface C. Also, remember, if applied MMC is as shown and if the feature is made to the LMC size limits of 12.5 diameter, there will be a bonus tolerance to add to the problem wall thickness. So apply this technique carefully.

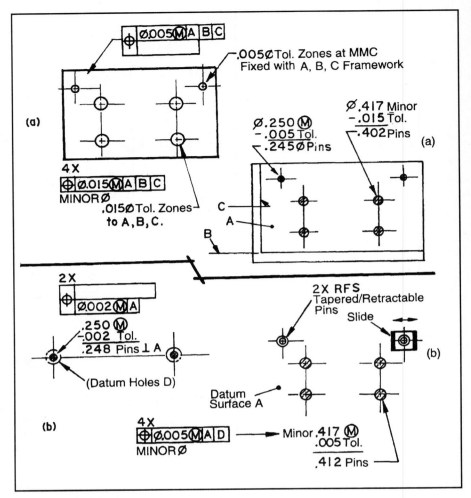

Figure 7-17.   Gaging impacts—multi datum controls for Figure 7-16.

Radial Hole Patterns

Radial hole patterns are common in cylindrical parts. Often the primary datum is the axis as seen in Figure 7-19a. The illustration shows a requirement at MMC which would allow a functional receiver gage to be developed. The gage would pass into datum diameter A and stop on datum surface B with six pins passing through the holes and into the gage. The depth of the holes in the gage would be equivalent to the wall thickness of the part.

Multiple Operations

It is common for more than one feature to share the same axis, such as holes, counterbores, or counter drills. If the feature control callout appears at the end of a series of operations in the note format, the implication is that the control specified is for all pre-

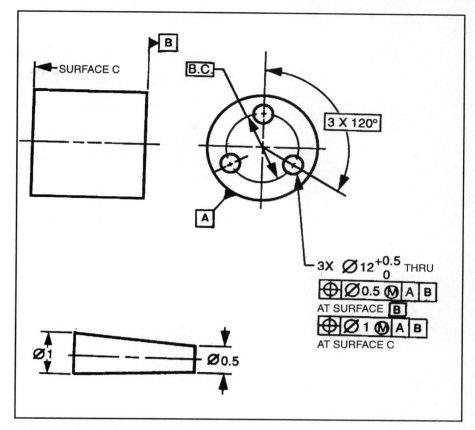

Figure 7-18. Position—conical tolerance zone.

viously listed feature operations. If this does not follow the design intent or requirements separate control frames should be specified after each feature specification as shown in Figure 7-19b.

Individual Holes as Datums

Some designs using shoulder bolts or ring dowels require each individual hole to serve as the datum for each counterbore. When this specification is required the design control callouts shown in Figure 7-20 may be used. The illustration shows the hole tolerance zone to be zero MMC, however, other tolerancing may be used. With shoulder bolts, we would have a fixed fastener condition. The tolerancing formula is $T = \dfrac{H - F}{2}$.

Datum Hole Tolerance
    Hole MMC = 20mm
    Bolt MMC = 20mm
    _____
    Available tolerance = 0

Counterbore Feature Tolerance
Counterbore MMC = 0.25mm
Bolt MMC = 24mm

Available tolerance = 1.0mm

1.0 ÷ 2 = 0.5mm each part

Other examples of individual hole datum features are illustrated in Figure 7-21 as applied at MMC. Note that the gage tends to resemble the mating part at closest fit conditions.

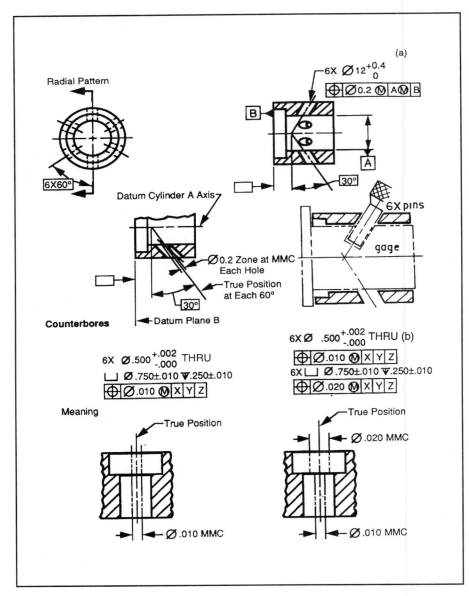

Figure 7-19.   Position—radial pattern and counterbores.

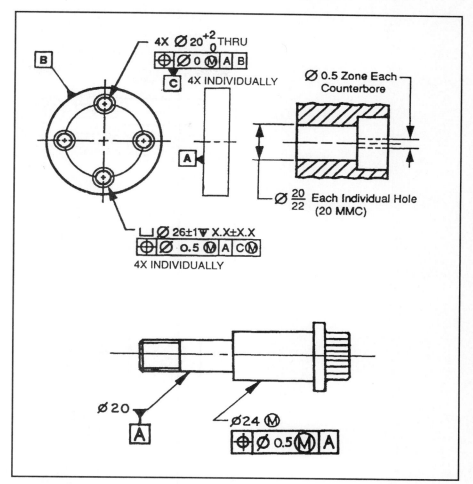

Figure 7-20. Position—hole alignment.

Other Tolerance Zones

Elongated Zones. Figure 7-22a illustrates the use of position tolerance and ± angular dimensioning to develop a banana-shaped tolerance zone. An angularity tolerance callout also could have been used with a slightly varied tolerance zone.

Restricted Zones. Figure 7-22b illustrates one method of restricting the tolerance zone to avoid thin walls at the three bosses. On occasion, the principles of MMC could work against us with this type of design. Figure 7-22b shows a zero tolerance at MMC. If the three holes are produced to the maximum limit size (LMC) of 13mm, the position tolerance would be a diameter of 1.0mm. This combination of the largest hole at the bonus tolerance limit would result in inadequate wall thickness (resultant condition). One method of avoiding this thin wall is through the

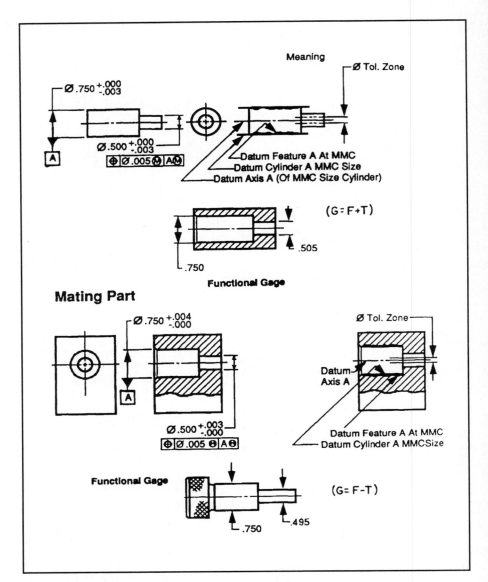

Figure 7-21.   Position tolerance—coaxial features.

use of a note as shown in the figure. The use of LMC in the feature control frame is also an option. (See Figures 7-22 and 7-40b.) As the holes in Figure 7-22b approach 13mm (LMC), their location must return to the true center in order to maintain the 6.0mm minimum wall specification.

Bidirectional Tolerance Zones

The technique illustrated in Figure 7-23 is used to create a bidirectional tolerance zone or rectangular zone. Note the absence

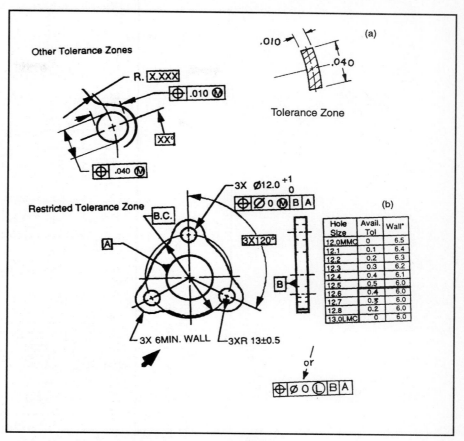

Figure 7-22. Restricted and other tolerance zones.

of the diameter symbol. Since round gage pins in round holes do not create rectangular tolerances, we might use the technique of designing the tolerance into the gage plate as shown. The pin diameter is equal to MMC, with the addition of a square shoulder. The tolerance is added to the shoulder size which then becomes the size of the rectangular hole in the plate. The rails of the gage simulate the secondary and tertiary datum edges B and C. The gage pins must pass through the gage plate and into the .365 MMC holes when the part is placed on the gage.

Tolerance Zones—Slots (Elongated Holes)

Slotted holes incorporate the *boundary principle*. This principle establishes a tolerance zone boundary when using positional tolerances. Position is normally applied to feature axes or centerplanes. By using the boundary principle in Figure 7-24a, not only are the tolerance zones well defined, but functional gaging is possible. The tolerance zone is the boundary around the shape of the slot. Figure 7-24 illustrates only one method for dimensioning slots. ASME Y14.5M-1994 provides other methods from which to choose.

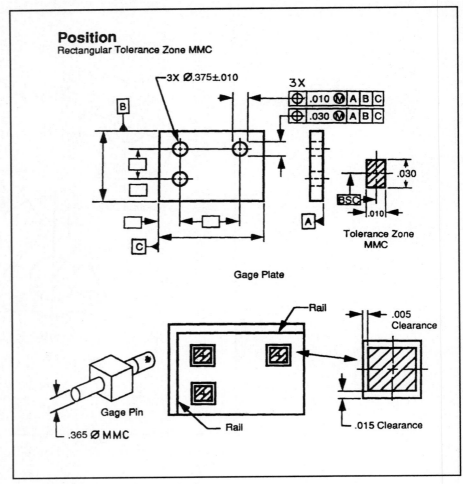

Figure 7-23.   Non-cylindrical (bidirectional) tolerance zones.

If the word "boundary" is not used, the tolerance zone applies at the intersection of the basic dimensions, and is rectangular as illustrated in Figure 7-24b. Functional gaging would be possible, however, the tolerance would be transferred to the gage plate, similar to that shown in Figure 7-23.

Composite Position Tolerance

Composite position tolerancing provides a composite application of positional tolerancing for the location of feature patterns as well as the interrelation (position and orientation) of features within these patterns. The position symbol is entered once and is applicable to both horizontal entries. Each complete horizontal entry in the feature control frame may be separately verifiable. (See Figure 7-25.)

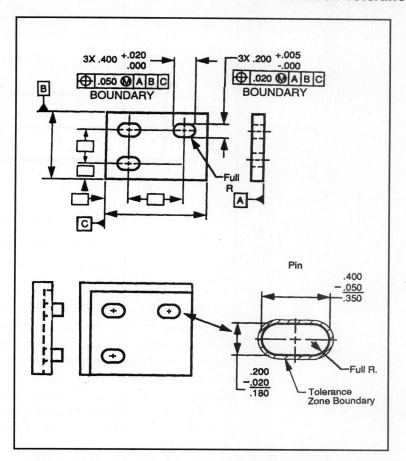

Figure 7-24a.   Boundary principle on elongated holes (slots).

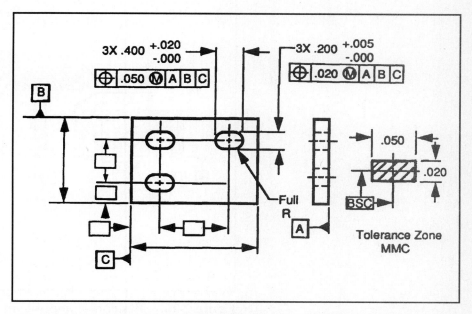

Figure 7-24b.   Position—elongated holes (slots) (rectangular tolerance zone).

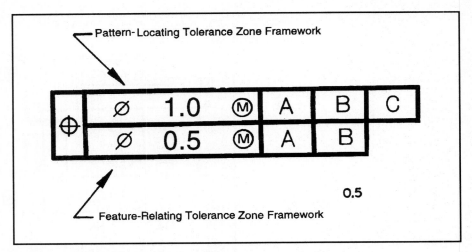

Figure 7-25. Composite position tolerance.

Where datum references are specified in the lower entry of the composite feature control frame, they govern the *orientation* of the feature-relating tolerance zone framework relative to the pattern-locating tolerance zone framework. One or more of the datums specified in the upper segment are repeated, as applicable and in the same order of precedence.

Figure 7-26 illustrates a composite positional tolerance control callout. The top portion of the callout specifies the four-hole

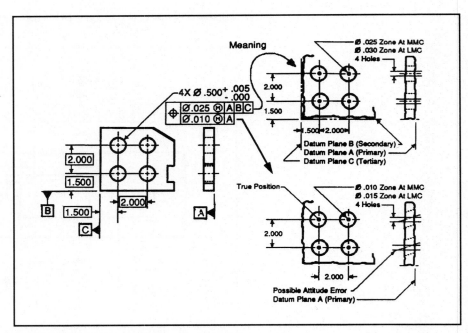

Figure 7-26.   Position—composite example.

pattern to be located from the datum framework A, B, C within a tolerance diameter of .025 at MMC. The lower callout specifies the individual holes to be relative to one another and perpendicular to datum A within a diameter tolerance of .008 at MMC. The upper callout is the hole pattern relative to the part, while the lower callout is the individual hole-to-hole tolerance. Two gages would be required.

Composite Position Tolerance Applications

Figures 7-27 and 7-28 illustrate the use of composite position tolerancing for the proper alignment of holes. As before, the upper callout controls the pattern of three holes relative to the datum framework A, B, C as a group, while the lower callout controls the hole-to-hole relationship and orientation (parallelism) to the respective datums as they are invoked.

In order to refine the orientation of the feature-relating tolerance zone governed by the boundary established by the pattern-locating tolerance zone, datum references are specified in the upper part of the frame and repeated as applicable. Do the same in the lower segment of the feature control frame.

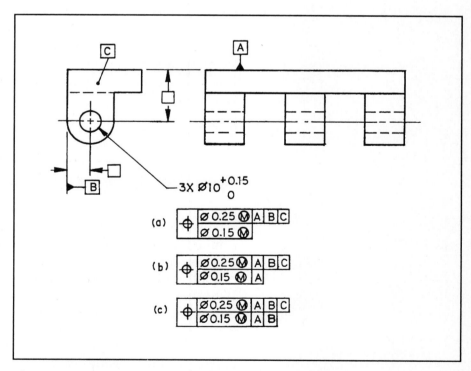

Figure 7-27. Composite tolerancing—hole alignment.

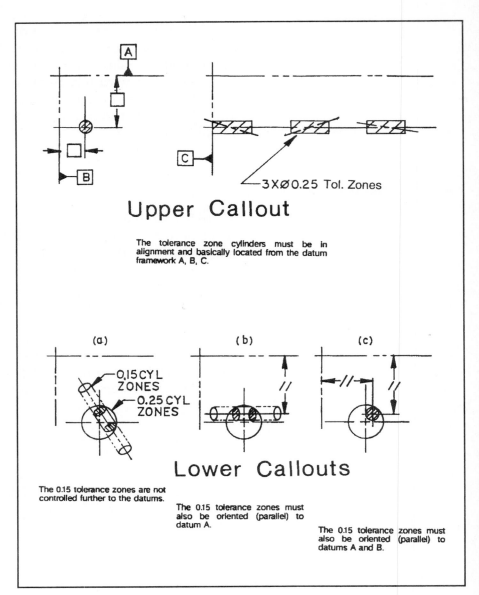

Upper Callout

The tolerance zone cylinders must be in alignment and basically located from the datum framework A, B, C.

Lower Callouts

Figure 7-28.   Upper and lower callouts of Figure 7-27.

With composite position tolerancing, it is important to remember that the *basic dimensions* that control the location of features *apply to the upper control callout only.* The lower control callout applies *orientation only*, relative to any datum references. If both orientation and location must be controlled to varied datums, multiple single segment position controls should be used as illustrated in Figure 7-39.

It is important to note that this principle also applies to composite profile tolerance controls as well. Comparisons of both composite position tolerancing and single segment position tolerancing are illustrated in Figure 7-36.

The upper callout in Figure 7-27 should be clear. The lower callout in Figure 7-27a has no datums invoked. Therefore, we have full movement or *full degrees of freedom* for the 0.15 tolerances zones to move, shift, and tilt within the larger 0.25 diameter tolerance zones. (See Figure 7-28a.) In Figure 7-27b, datum A is invoked removing one degree of freedom of movement. Therefore, the 0.15 diameter tolerance zones must also be parallel to datum A as shown in Figure 7-28b. In Figure 7-29, datums A and B have been invoked, removing two degrees of freedom. This means that the 0.15 diameter tolerance zones must also be parallel to datum B. This condition results in (three) 0.15 diameter aligned and oriented tolerance zones which are free to float within the larger 0.25 diameter zones, but which must be parallel to both datums A and B. (See Figure 7-28c.) Since datum C is not specified in the lower callout, it is not invoked.

Composite Circular Hole Patterns

Figure 7-29 illustrates the use of composite tolerancing on a circular pattern of holes located from a primary datum surface G and a secondary datum diameter H at MMC. The same principles apply as before. That is that the upper callout controls the pattern of six holes relative to the datum framework, while the lower callout controls the hole-to-hole tolerance and perpendicularity to datum G.

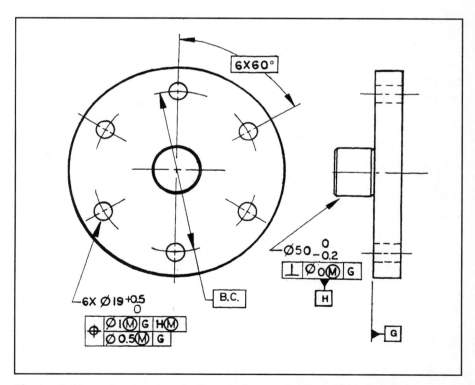

Figure 7-29.   Composite position tolerancing–circular pattern.

Functional Gage-Part at MMC

Using gaging formulas, the gage pins for the upper callout in Figure 7-30 are 18.0 diameter (19.0 MMC minus a tolerance of 1.0), which must be relative to primary datum G. The gage must fit over the secondary datum H at its virtual condition (perpendicular to G with a zero tolerance at MMC). A second gage is required. It must have pins 18.5 in diameter (19.0 minus 0.5 at MMC), but without the center datum H invoked. This second gage requires closer hole-to-hole tolerance and perpendicularity to G.

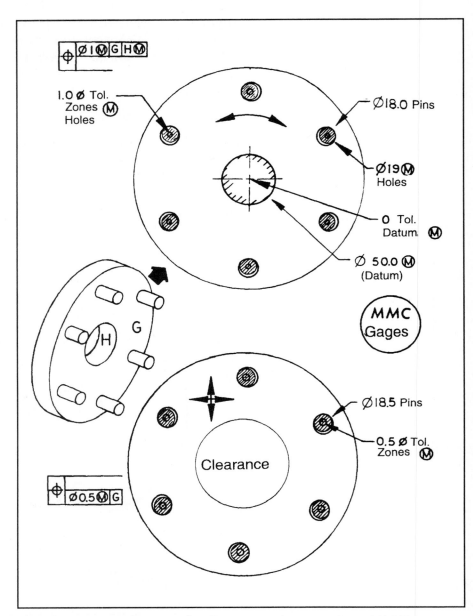

Figure 7-30.   Explanation for Figure 7-29 (part made to MMC).

By being centered on the datum H, the top gage in Figure 7-30 would allow partial radial movement only. The gage in the lower portion of Figure 7-30 would allow multi-directional movement, but of a lesser amount. Figure 7-30 illustrates a part made to MMC specifications with both gages represented.

Figure 7-31a illustrates a part made to LMC specifications (largest holes) with a gage shown for checking the upper callout. This figure shows the effects of bonus tolerance on both the feature holes and the datum H. Effects of LMC and bonus tolerance for the feature holes in the lower callout would be similar, but without the impact of secondary datum H.

Figure 7-31b illustrates the effect of LMC holes and bonus tolerances with a possible impact on minimum wall thickness. Always consider this resultant condition when applying tolerances at MMC. A comparison of composite position tolerancing and single segment tolerancing of circular hole patterns is illustrated in Figure 7-35.

Figure 7-32 illustrates a possible functional gage when features are specified MMC but the datum hole H is specified RFS. Split rings or other variable gage features for the datum hole would be required.

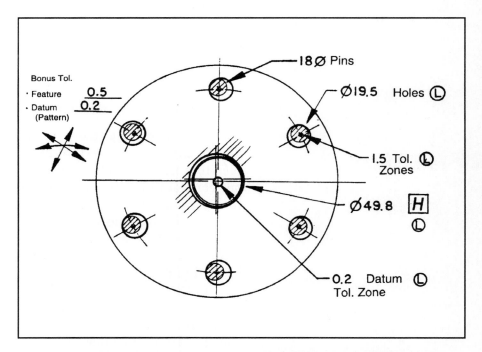

Note: A functional gage accepts a 0.2 bonus tolerance at the datum axis due to the datum at LMC, and a 0.5 bonus tolerance at each feature axis due to the feature at LMC and not 0.7 tolerance at each feature axis. Remember this when doing open set-up or CMM inspections. (See paper-gaging section and Figures 8-11 and 8-12.)

Figure 7-31a.   Explanation of Figure 7-29 (part made to LMC).

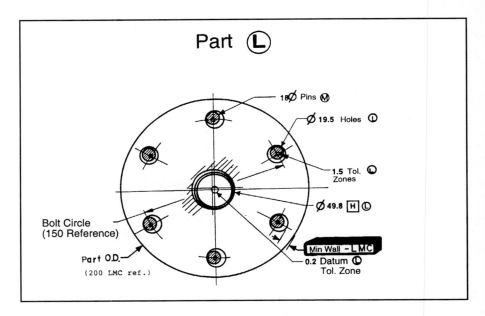

Note: A minimum wall could evolve when both features and datum(s) are LMC. Minimum walls may be determined as follows:

O.D. at LMC (200) – B.C. (bolt circle 150 ref.)
= 50 – LMC (6) holes (19.5)
    – (6) holes loc. tol. to datum H LMC (1.5)
    – datum H size tol. (0.2)
    – datum H loc. tol. to O.D. (if given)

50 – 21.2 = 28.8 + 2 = 14.4
Minimum wall = 14.4 LMC

Figure 7-31b. Explanation of Figure 7-29 (part made LMC; effect on minimum wall).

Radial Hole Patterns (Datum Axis Primary)

When locating a pattern of holes or features from a primary datum axis, we are extending the composite tolerancing principles to a more complex degree. (See Figure 7-33.) The upper callout control illustrates a cylindrical part with radial holes relative to a primary datum axis, a secondary datum face, and a tertiary datum keyslot. The illustration shows the four holes relative to the datum framework within a 0.8 diameter tolerance zone at MMC. We can see that the axis of the 0.8 zones basically must be located and oriented to the datum framework.

Let's work through the lower callout, starting with no datums invoked. Figure 7-33a shows the hole-to-hole tolerance zones to be a diameter of 0.2 and that these zones must lie within the 0.8 diameter zones of the upper callout. The 0.2 diameter zones have no datums invoked and, therefore, the 0.2 zones have full degrees of freedom within the 0.8 zones (slide, tilt, and rotate). In Figure 7-33b, we have invoked datum A (axis) as the primary datum, which allows the 0.2 hole tolerance zones to slide

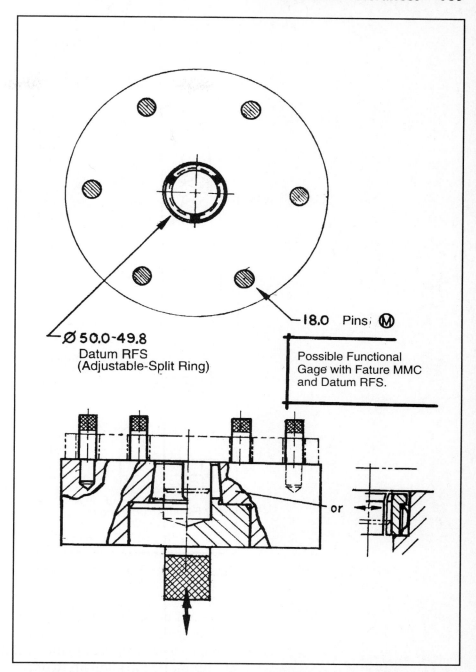

Figure 7-32.   Possible functional gage with feature MMC and datum RFS.

and rotate as long as they remain within the 0.8 diameter zones and are oriented (perpendicular) to the primary datum axis A.

In Figure 7-33c, we have invoked secondary datum surface B which requires that the four holes' axes be oriented (parallel) to datum B. Datums are defined as mutually perpendicular, however, we remember we cannot make perfect parts, and there will be

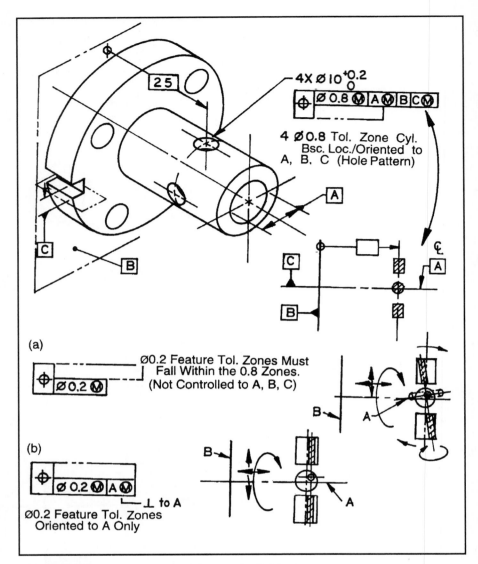

Figure 7-33.   Datum axis primary, radial holes.

some out of perpendicularity error between datums A and B. This will be reflected in the secondary datum B. By controlling the holes perpendicular to primary datum A, we will achieve a degree of automatic parallelism to datum B, within the allowable square-ness limits imposed between datums A and B.

Figure 7-33d has added the third datum keyslot C, which stops the rotation of the four holes. The 0.2 tolerance zones must be perpendicular to datum A at MMC, parallel to datum B, with the rotation of the tolerance zones stopped by datum C at MMC. If we look at the completed callout, we see the upper and lower datum controls invoked as identical. This may appear to make the upper callout redundant, however, as the lower callout

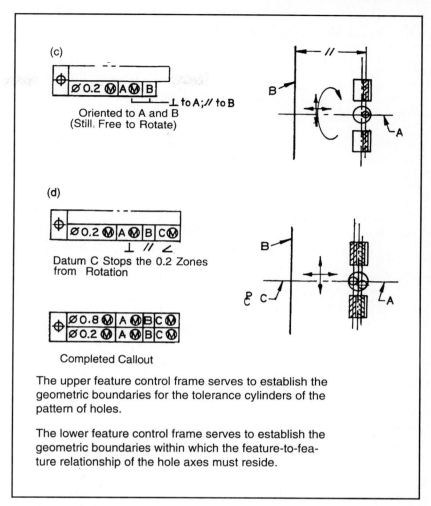

Figure 7-33 *(cont.)*.

controls *orientation only*, and as we have worked through the requirements of both control callouts, we can see the use for both.

These conditions and circumstances are unique to cylindrical type parts with a primary datum axis where the orientation-only definition is applied. Designs should be carefully thought through before applying datum frameworks to these types of parts as this technique tends to complicate composite principles. The logic of the datum controls and design requirements must be clear in order for the user to understand our intent. Remember, if a datum is not wanted, do not invoke it. If it is used in the lower control frame, it is applied as *orientation only*! A comparison of composite position tolerancing of radial hole patterns and single segment tolerancing is shown in Figures 7-37 and 7-38.

Composite Versus Multi-Single Segment Position Tolerancing

Let's stop a moment and compare composite and multi-single segment position tolerancing to help clarify the two concepts. *Single segment position callouts* control the *location and orientation* of feature relationships to the datum framework. The use of multiple callouts does not change this concept, they simply add or delete datums and associated controls.

Composite position tolerancing, per the ASME Y14.5-1994 standard, invokes the *orientation only* principle and uses one position tolerancing symbol for two distinct purposes. The upper (top) control callout controls the feature or pattern of features to the datum framework. The lower callout controls the feature-to-feature relationship and the orientation of the pattern to the datums specified. The lower callout has no relationship to the basic dimensions of pattern location and, therefore, locationally need only fall within the upper control tolerance zones.

To make a comparison of these concepts, we will use Figure 7-7a with a rectangular hole pattern, Figure 7-27 with an aligned hole pattern, Figure 7-29 with a circular hole pattern, and Figure 7-33 with a radial hole pattern as examples.

In Figures 7-34 through 7-38, all callouts are MMC for the features and RFS for the datums. There could be bonus tolerances for the features and/or datums based upon size variation, however, we will concentrate on the one concept for now. Compare the concepts, controls, datums and symbols used and note the results, using both types of controls.

Least Material Condition (LMC)

Positional tolerances may be applied with the features specified at LMC. For holes, this is a condition where the holes are the largest diameter within the size tolerance. The most common application of this callout is to control wall thickness. MMC allows holes to be located with bonus tolerances as the holes increase in size. This MMC application can result in inadequate wall thickness when holes are close to bore diameters or part edges. With LMC, the control is applied when the hole is at the largest possible size limit. Bonus tolerance is added as the feature departs from LMC. (See Figure 7-40.)

Concentricity

Concentricity is the condition where the *median points* of all *diametrically opposed* elements of a feature's surface of revolution

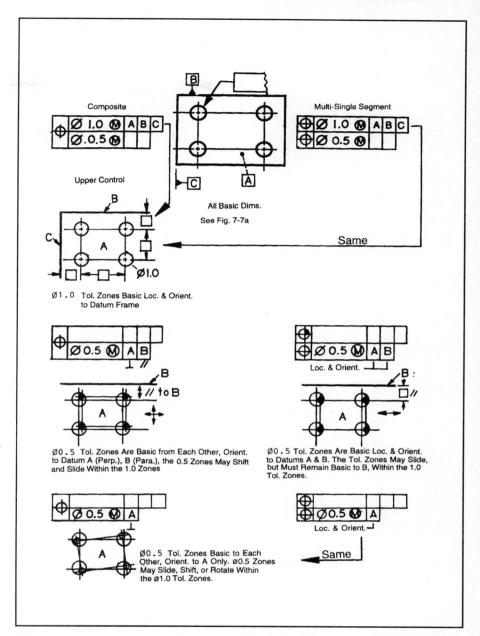

Figure 7-34. Composite versus multi-single segment position tolerancing for Figure 7-7.

(such as cylinders, cones, spheres, hexagons, etc.) are coaxial with the axis of a datum feature. A *concentricity tolerance* is a cylindrical (or spherical) tolerance zone whose axis (or center point) coincides with the axis (or center point) of the datum feature(s). The median points of all corresponding elements of the feature(s) are controlled, regardless of feature size, and must lie within the cylindrical (or spherical) tolerance zones.

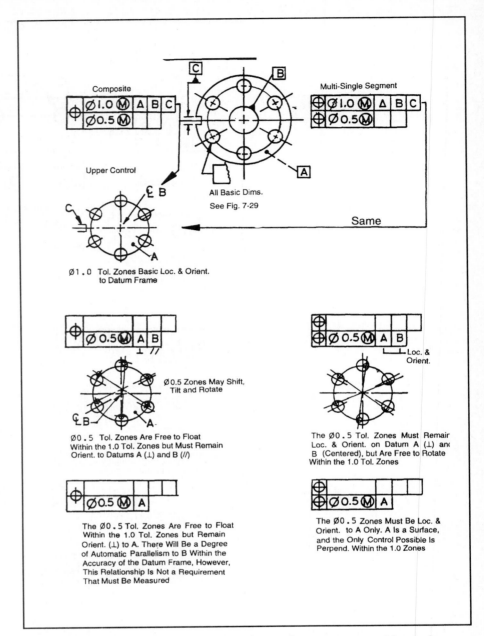

Figure 7-35. Composite versus multi-single segment position tolerancing for Figure 7-29.

Concentricity is considered a location control, in that it controls a feature axis relative to a datum feature axis. The formal definition uses the terms "a feature median line relative to a datum feature axis." The median line must be derived from surface elements, as we will find out through the use of these exercises. Concentricity is always RFS and considered difficult to measure because the tolerance zone exists at the feature axis. (See Figure 7-41.)

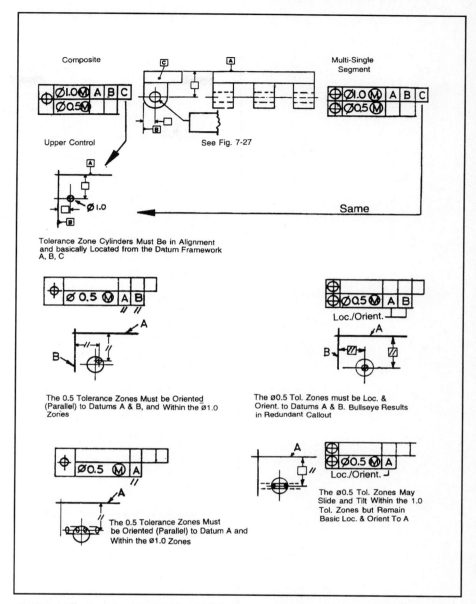

Figure 7-36. Composite versus multi-single segment position tolerancing for Figure 7-27.

Since the tolerance zone is at the feature (median line) axis, we must derive this median line tolerance zone by using the feature surface. Figure 7-42a illustrates the use of a single indicator device as an example for obtaining measurements. With a single indicator, it is difficult to determine between form (circularity) and location error. Without further assessment, we do not know if the feature is not concentric or not round. Further, a feature may be within size limits but be out of round. It could be rejected for the wrong reasons.

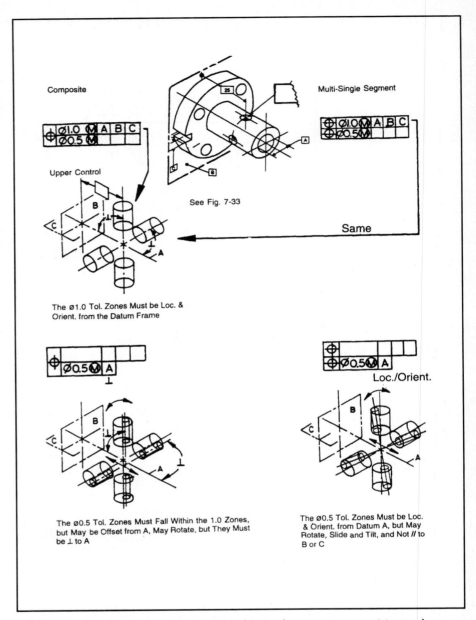

Figure 7-37.   Composite versus multi-single segment position tolerancing for Figure 7-33.

A common method for concentricity measurement is the use of two indicators in opposed locations. Using the formula shown in Figure 7-42b, we can plot a locus of points for one full revolution of the part about the datum axis. "X" is the datum location in the measurement set-up and may be any known quantity. As we move along the measured feature, we connect the plots of points to determine the concentricity of the feature's median

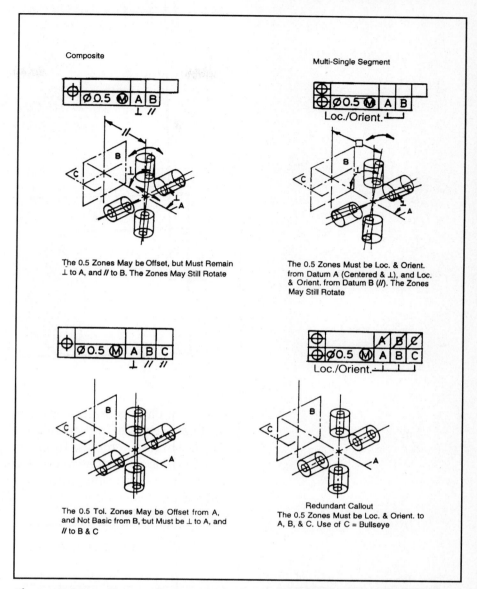

Figure 7-38. Composite versus multi-single segment position tolerancing for Figure 7-33.

points (a cylinder zone). This method works well except for shafts with an odd number of lobes (i.e., 3 through 19). A computer program would be helpful in determining the axis in these instances.

More advanced techniques also will do the job, including air or electronic gaging, computer-assisted measurement, and high-resolution magnification. These methods, though, are more complex and usually more costly.

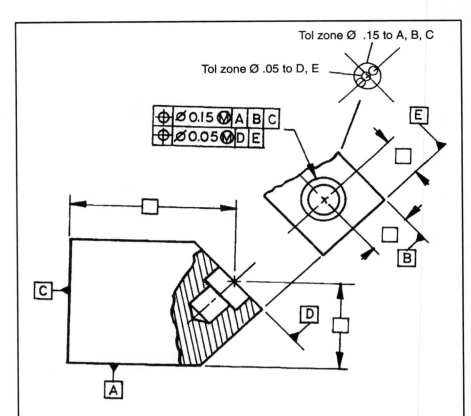

Figure 7-39.   Multiple single segment control frames.

Datum Error and Concentricity

By definition, concentricity is relative to a datum axis! In evaluating concentricity, we must be aware of the impact of vee locators and stulus location as a datum set-up. *Datum error* may be read at the feature surface being measured, which could effect a part acceptance decision. The same conditions that effected runout reading decisions will also impact concentricity if evaluated in the same manner. Figures 7-43 and 7-44 remind us of the results of datum error, vee locator type, and indicator(s) position. Once again, the inspection set-up is critical in order to achieve repeatable and accurate measurement results.

Because of the complexity of finding the derived median line of a feature and with all the influential variables involved, it is

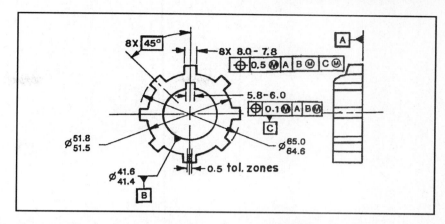

Figure 7-40a. Positional tolerance of noncylindrical features—tabs.

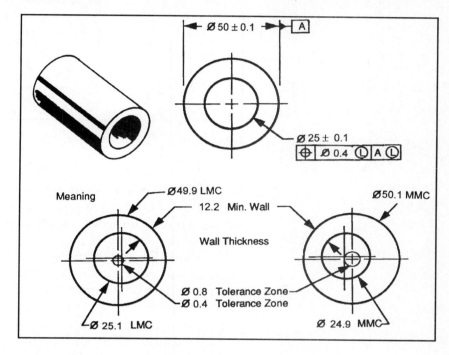

Figure 7-40b. Position—least material condition.

easy to understand why position or runout might be preferred over concentricity in a production or factory environment. By it's definition, concentricity is more suited to a lab environment or if very sophisticated equipment is available.

Selection of Coaxial Controls

Coaxiality is that condition where the axes of two or more surfaces of revolution are coincident. The amount of permissible vari-

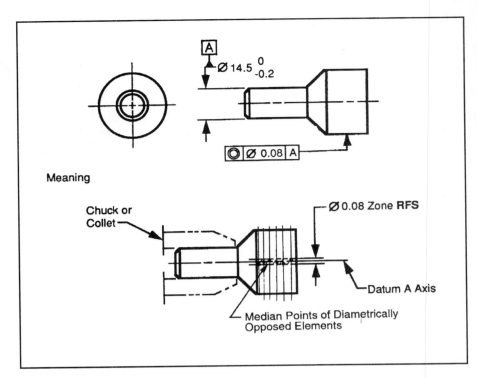

Figure 7-41. Concentricity.

ation from coaxiality may be expressed by a position tolerance, a runout tolerance, or a concentricity tolerance. Selection of the proper control depends on the nature of the functional requirements of the design.

Generally, a design configuration logically will help determine which coaxial control to use. Figure 7-45 illustrates some key reminders which may also help. They include:

- *Runout*—Runout is a composite error read at the feature surface (always RFS) and acceptable when other controls of size and/or form are in place. Circular or total runout controls are usually available. However, total runout requires movement of the indicator parallel or perpendicular to the datum axis (or surface).

- *Position*—Position controls normally apply to a feature axis or centerplane but may be applied RFS, MMC, or LMC with MMC and LMC providing bonus tolerance. MMC also provides the ability to use functional gaging, if desired. Position tolerance is generally applied to static fit assemblies and hole patterns. It creates a virtual mating envelope when applied at MMC.

- *Concentricity*—Concentricity is not commonly used as it applies to a feature median line (centerline) relative to a datum axis (both at RFS). Because of the reasons just covered in the con-

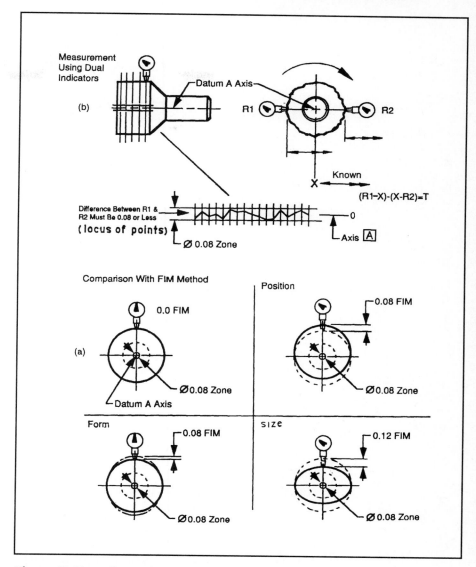

Figure 7-42.   Concentricity.

centricity section, runout or position tolerances are generally recommended over concentricity. Concentricity historically has been used when it was necessary to control the axis of a rotating mass to a datum.

Symmetry

Symmetry is that condition where the median points of all opposed or corresponding elements of two or more feature surfaces are congruent with the axis or center plane of a datum feature.

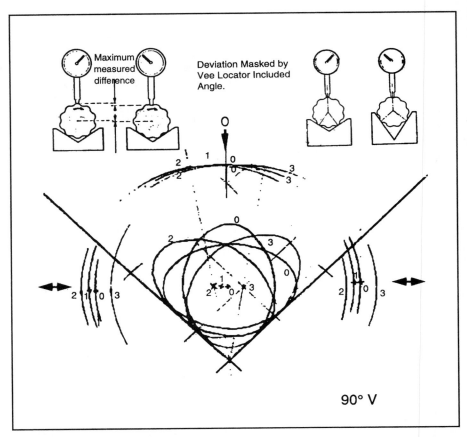

Figure 7-43.   Datum ovality—measured error with 90° vee locators.

Symmetry controls were removed from the ANSI Y14.5M-1982 standard because it was felt that we could achieve the same results through the use of true position RFS. This, however, was found not to be the case. Symmetry and concentricity controls are similar concepts and deal with the location of median lines or points. But the concepts do not deal with maximum inscribed or minimum circumscribed cylinders, which are the basis of position controls.

This change left the U.S. standard and ISO standard out of sync since ISO continued the use of symmetry controls. Many persons felt with the mathmatization of the standard, and due to the preciseness of the definitions used, it was necessary to return symmetry as a precisely defined control of a feature median plane to a datum centerplane. (See Figure 7-46.)

With the addition of a symmetry definition and symbol to the ASME Y14.5M-1994 standard, compatability between the U.S. and ISO standards returns.

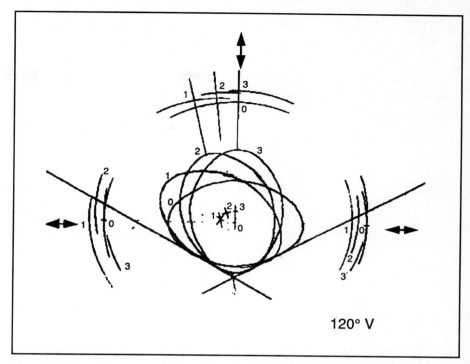

Figure 7-44.   Datum ovality—measured error with 120° vee locators.

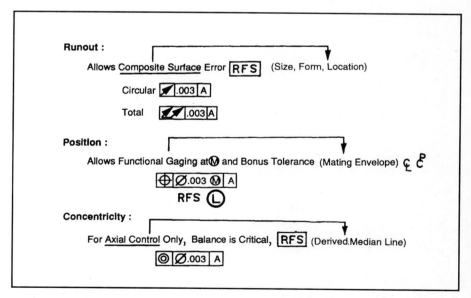

Figure 7-45.   Selection of proper control coaxial features.

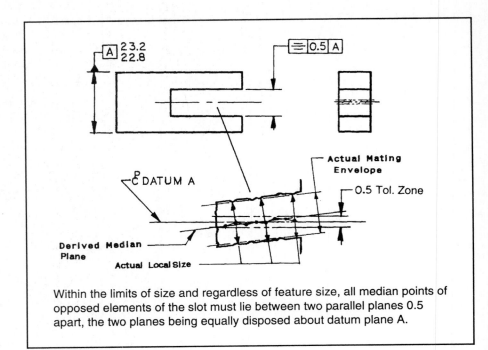

Within the limits of size and regardless of feature size, all median points of opposed elements of the slot must lie between two parallel planes 0.5 apart, the two planes being equally disposed about datum plane A.

Figure 7-46.   Symmetry (RFS).

EXERCISE 7-1.   POSITION TOLERANCE

1. True position controls are implied MMC.   T   F
2. A true position tolerance zone is a ± zone.   T   F
3. (M) allows a bonus tolerance.   T   F
4. True position controls require use of datums for hole *pattern* controls.   T   F
5. Indicate the correct position tolerance associated with each of the following hole sizes for the object shown below:

Size	Tolerance
8.4	
8.5	
8.6	

6. The virtual condition for this hole is _____ .

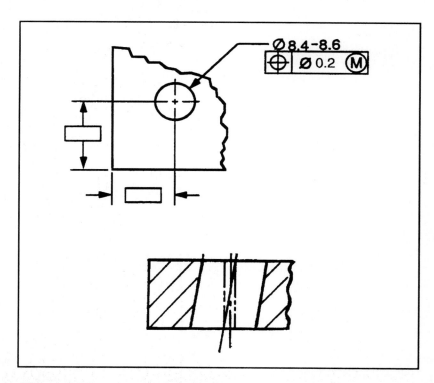

Figure 7-47.

EXERCISE 7-2. FORMULAS

What are the formulas for:

1. Floating fastener position tolerance.

2. Fixed fastener position tolerance.

3. Hole diameter virtual condition at MMC.

4. Shaft diameter virtual condition at MMC.

EXERCISE 7-3.   APPLYING THE FORMULAS

The design in Figure 7-48 is a shaft with an assembled key and a mating gear which must mount to the shaft, over the key, and be secured with four M10 bolts. Given your formulas from Exercise 7-2 (and page 141), complete the following:

1. Determine the feature controls, datum framework, and tolerances of the design. Apply 60% of the available tolerance to the shaft and key assembly, and 40% to the gear. Apply the tolerances at MMC of the features and datums. Also determine the maximum hole size for the gear holes.

2. Based on these controls and tolerances, complete the gage designs for Figures 7-49 and 7-50 by filling in the missing values. Ignore gaging tolerances for now.

In determining the datums, it may help to visualize the most influential features of the design and apply the diameter length to surface area ratio. Also, since this is a static fit application, and since we are to design gages to MMC conditions, it is logical to use the same datum sequences and surfaces on both parts.

Review the exercise in Figure 7-51. To help in getting started, either break each surface relationship down to the basic formula $T = \dfrac{H - F}{2}$, or use the summary formula approach as follows:

[Sum of Female Feature Diameters @ MMC] – [Sum of Male Feature Diameters @ MMC] = Sum of Tolerances

This value is the total tolerance to be applied to all features as required. Each feature controlled in Figure 7-51 has a virtual condition (mating size envelope). Determine those values.

T = H - F (60/40)

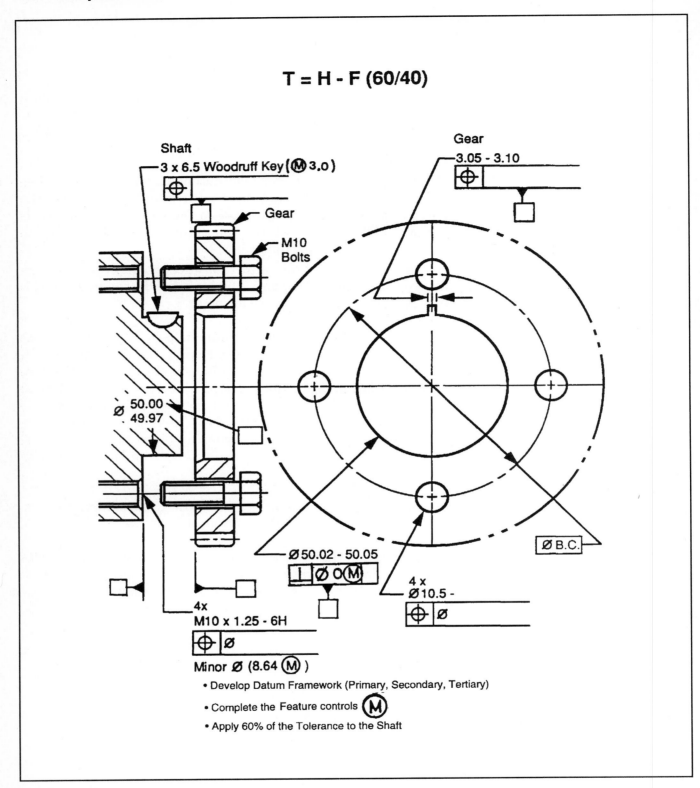

Shaft
3 x 6.5 Woodruff Key (Ⓜ 3.0)

Gear

M10 Bolts

Gear
3.05 - 3.10

Ø 50.00
49.97

Ø 50.02 - 50.05

⊥ | Ø 0 Ⓜ |

Ø B.C.

4 x
Ø 10.5 -

⊕ | Ø

4x
M10 x 1.25 - 6H

⊕ | Ø

Minor Ø (8.64 Ⓜ)

- Develop Datum Framework (Primary, Secondary, Tertiary)
- Complete the Feature controls Ⓜ
- Apply 60% of the Tolerance to the Shaft

Figure 7-48.

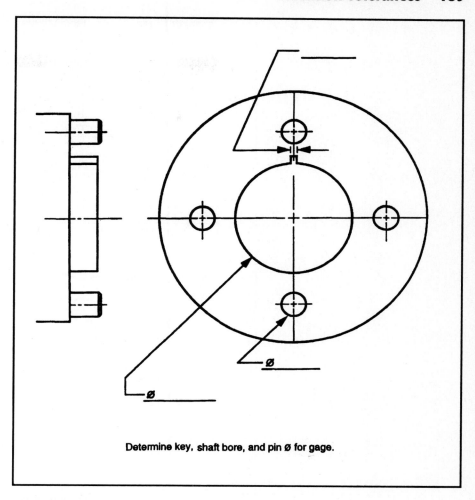

Determine key, shaft bore, and pin Ø for gage.

Figure 7-49.   Gear gage.

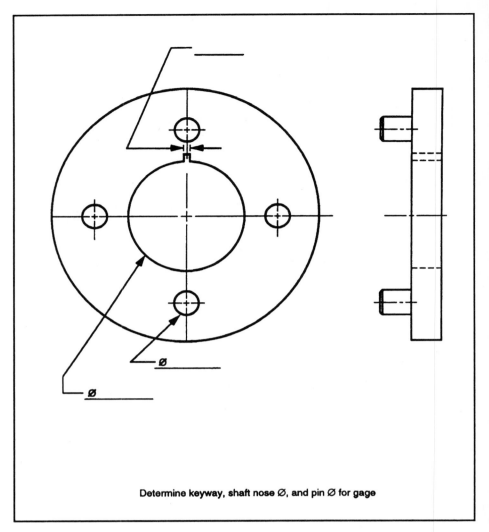

Determine keyway, shaft nose Ø, and pin Ø for gage

Figure 7-50. Shaft gage.

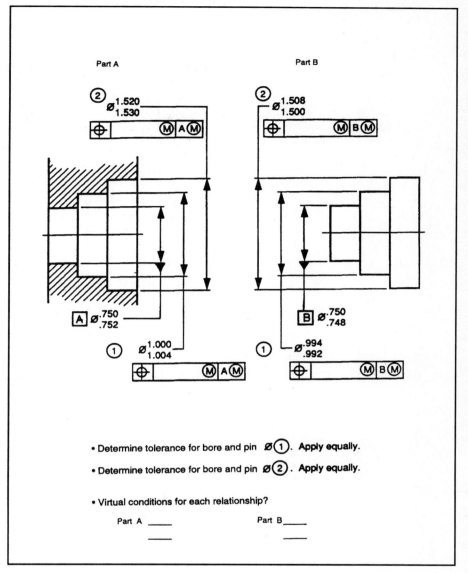

Figure 7-51.

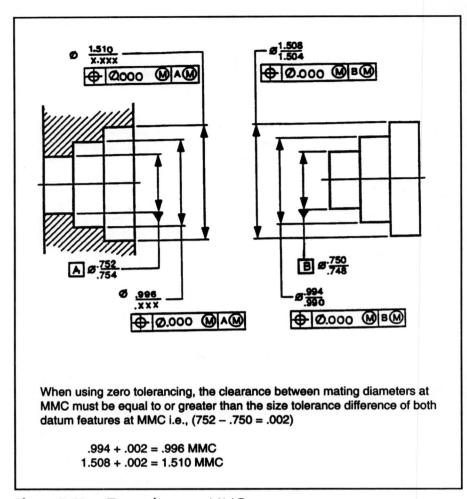

Figure 7-52.   Zero tolerance—MMC.

Figures 7-53 through 7-58 represent a simple part with a center hole and a pattern of four smaller holes. Follow the directions on each figure, keeping in mind when datum surface B is used as a tertiary datum feature along with secondary datum feature hole RFS, surface B will be adjustable in that it serves to establish orientation to the datum framework.

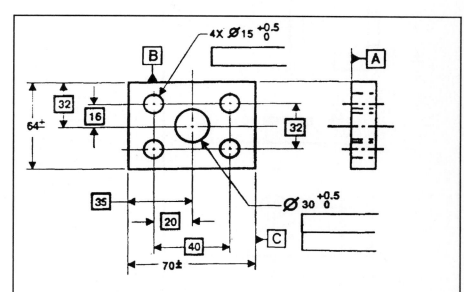

- Functional gaging is to be used for below requirements.
- The large dia. center hole is to be located from datum framework A, B, C within Ø 0.8.
- The large dia. center hole also to be perpendicular to datum surface A within Ø 0.5.
- Large hole to be datum D.
- The 4 small holes are to be relative to datum A and datum D at MMC within Ø 0.5 at MMC.
  1. Complete the feature control frames.
  2. Identify the tolerance zones and gage pin dias. in the views that follow.

Figure 7-53.

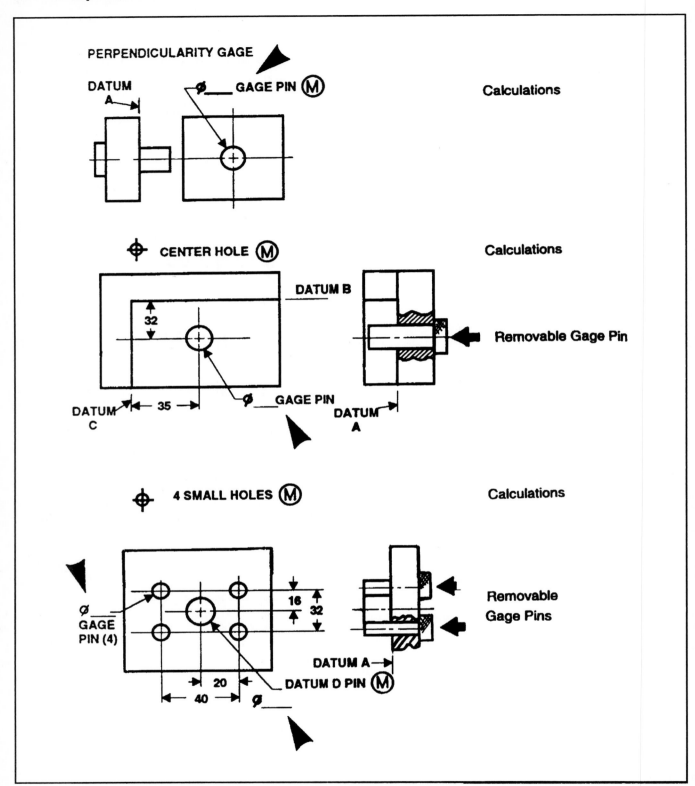

Figure 7-54.

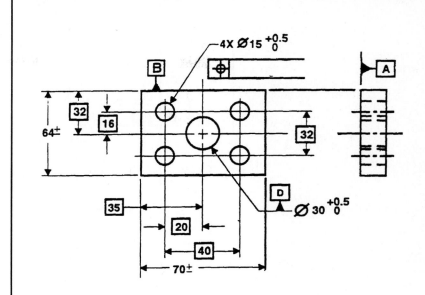

- Adjust the callout so the 4 small holes are MMC, but relative to the large hole RFS, and tertiary datum B.

- Tolerance zones are similar to the previous exercise except as noted.

Figure 7-55.

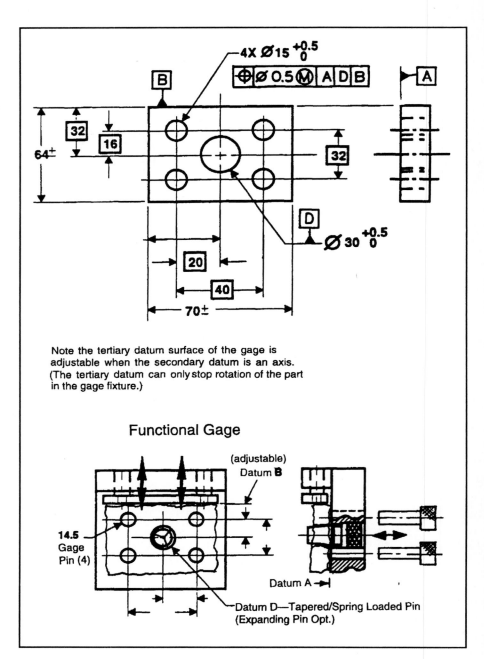

Note the tertiary datum surface of the gage is
adjustable when the secondary datum is an axis.
(The tertiary datum can only stop rotation of the part
in the gage fixture.)

Figure 7-56.

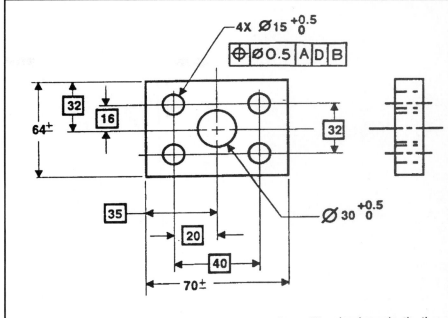

The principles of functional gaging cannot be utilized when both the datum and feature are specified RFS. Open set-up inspection, (CMM), or the techniques of paper gaging would be necessary.

Figure 7-57.

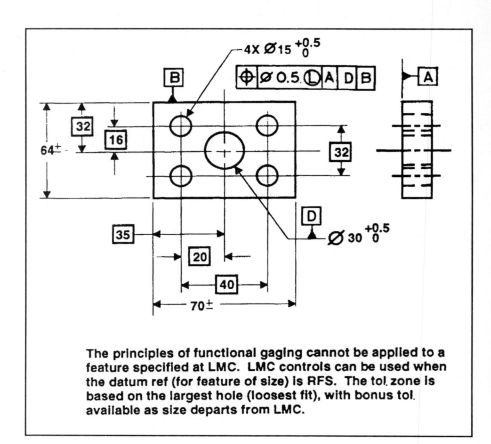

The principles of functional gaging cannot be applied to a feature specified at LMC. LMC controls can be used when the datum ref (for feature of size) is RFS. The tol. zone is based on the largest hole (loosest fit), with bonus tol. available as size departs from LMC.

Figure 7-58.

EXERCISE 7-4.   CONCENTRICITY

1. Concentricity is always datum related.   T   F
2. Concentricity is applied _____. (MMC, LMC, RFS)
3 The concentricity tolerance exists at the feature _____ and relates to the datum _____.
4. Therefore, the use of V-blocks for datum set-up is recommended.   T   F
5. Concentricity controls require the verification of feature axes, without regard to surface conditions, to datum axes. Therefore, unless there is an exact need for this control, the use of position or runout is recommended.   T   F
6. Indicate the middle and smaller diameters of the object in the figure below to be concentric with the larger diameter within 0.1. Draw the tolerance zone.

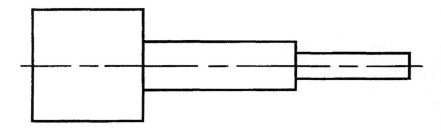

8

Paper Gaging

Paper gaging is the mathematical and graphical manipulation of inspection data derived by gaging means other than functional gaging. Now that we understand the formulas for tolerancing and know how to develop functional gages, we need to be able to use these concepts to accept parts, of the same quality, as the functional gage but with open set-up inspection or CMMs. Time constraints, volume, or gage expense may prohibit the use of dedicated gaging. Process control system development may require audits for feedback to the engineering and manufacturing areas. For these reasons, it is necessary to understand the principles behind *paper gaging analysis*. The examples that follow will walk you through a typical inspection process using paper gaging principles.

Before we get into paper gaging and position tolerancing, let's take a look at a coordinately (±) dimensioned part with a six-hole pattern. Figure 8-1 illustrates a hole pattern which is located from the part edges by the two 1.000± .015 dimensions. The hole-to-hole tolerance is 2.000± .010, and 4.000± .010.

To evaluate this part without functional gaging, we could first find out if the pattern (6 holes) fell within the pattern locating tolerance of 1.000± .015 (in both directions). (See Figure 7-3.) The second step could then be evaluation of the hole-to-hole tolerance (±.010). The most common method is to locate one hole (any corner hole) and then another hole some distance away (say, holes #1 and #3), thus establishing two 0-0 (or X-Y) reference planes. From this set-up, we take measurements and find the actual hole

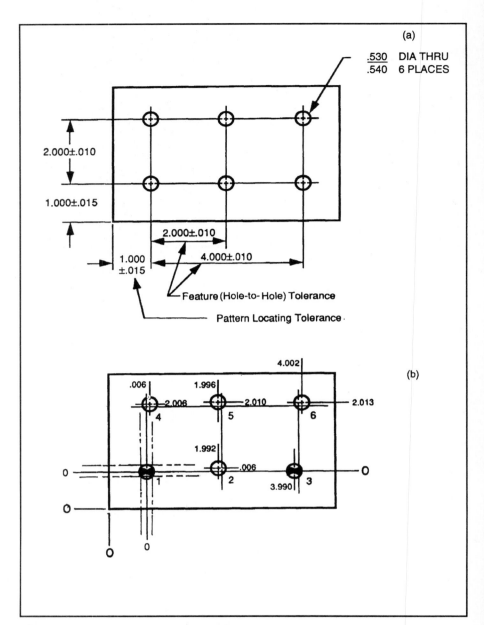

Figure 8-1. Paper gaging techniques—open set-up inspection.

locations of all the remaining holes in the pattern as shown in Figure 8-1b.

Figure 8-2 illustrates the holes that all lie within the ±.015 pattern locating tolerance as plotted on ¼-inch graph paper with a scale of one square=.002 inch (125:1). The center of the plot also serves as the center (perfect) of each hole in the pattern. Looking at the figure, try to answer these questions:

1. Do all holes lie within the hole-to-hole tolerance of ±.010?

2. Is the part within print tolerance and will it function as designed?

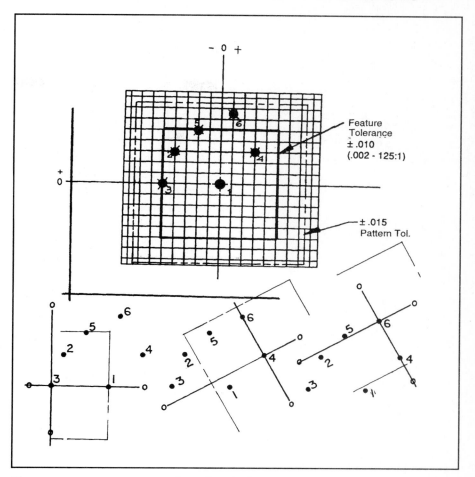

Figure 8-2. Paper gage plot of six-hole pattern.

Using coordinate methods of dimensioning and by giving one hole a value of zero, we would conclude that hole #6 is outside the tolerance limit and, therefore, the part is not to print tolerance and will not work.

Using this method, without further evaluation of the data, will most likely result in rejection of good parts. Let's examine it further. We will not always use the same hole(s) for our measurements. We may use holes #6 and #4 for establishing the planes. But we could use combinations of holes #3 and #1, and holes #4 and #6, which will yield different results. Yet the holes will remain the same in relation to one another. Using this method (making one hole zero) removes the ±.010 tolerance from the initial hole (#1), even though hole #1 is part of the pattern of six holes, and even though hole #1 could become hole #6 if another inspector were to measure the part.

Because we have no control over the holes selected, we must have some way of allowing tolerance for holes #1 and #3. We can do this through the use of a transparent overlay (to the same scale of .002:1). We can reevaluate the hole locations by adjusting the ±.010 overlay .002 left and .004 vertical to establish a new location for the tolerance zone (as long as we do not violate the ±.015 pattern locating zone).

This concept is illustrated in Figure 8-3. Let's look at it again. Do all holes lie within the hole-to-hole tolerance of ±.010? Is the part to print? Will it function as intended? The answer to these questions is "yes." Using the methods of two holes for inspection set-ups without giving proper consideration to the holes made zero for the establishment of measurement planes can reject acceptable parts and create unnecessary rework costs.

The process of paper gaging uses the same principle found in *bestfit centers* software in most CMM measurement devices. We may wish to look to our present measurement methods to see if set-ups and the decisions made about them are contributing to an excessive scrap rate, or to rework costs. This may also explain why apparent "bad" parts seem to fit at assembly.

With this approach and the tolerance method, all measurements are RFS. We have not taken advantage of the full tolerance, (circular zone), nor have we considered the bonus tolerance if the holes are larger than MMC. These benefits are not available to us with the ± system.

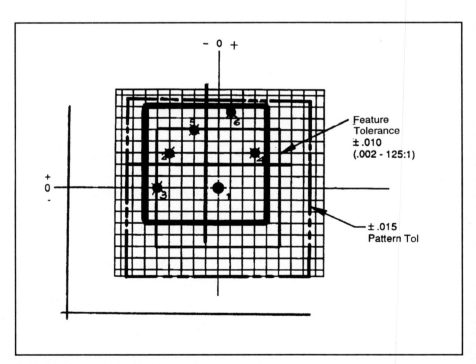

Figure 8-3.   Paper gage plot—adjusted bestfit pattern location.

Paper Gaging and Position Tolerancing

The part in Figure 8-4 is similar to that in Figure 8-3. It has been dimensioned using position controls per ASME Y14.5. The hole pattern is dimensioned and located from the part edges (datum reference frame) within a diameter tolerance of .020 at MMC. The hole-to-hole tolerance is a diameter tolerance of .012 at MMC and perpendicular to datum A. Using the same procedure as before, we locate from the datum feature surfaces in order to establish the pattern location.

Next, we need to measure the location of each hole to determine its departure from the true position and plot the holes on the graph. Further, we need to measure the size of each hole in order to apply the bonus tolerance that is available because of MMC.

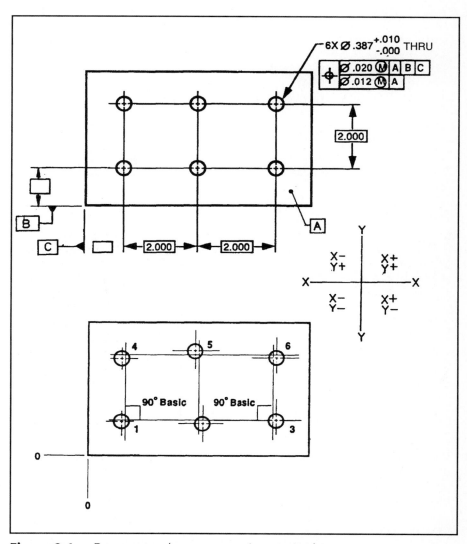

Figure 8-4.   Paper gage (open set-up inspection).

In the example in Figure 8-5, we've used graph paper and a scale of one ¼-inch square = .001. Plot the hole locations and determine the available tolerance for each hole by adding the amount of oversize to the position tolerance for each. The values are shown in Figure 8-5b.

With our overlay of the same scale, we can see that the holes all lie within the specified tolerance of .020 diameter at MMC. The center of the plot also serves as the center of the datum framework. Note that the available tolerance for the holes is larger than .020 because none of the holes were produced to MMC.

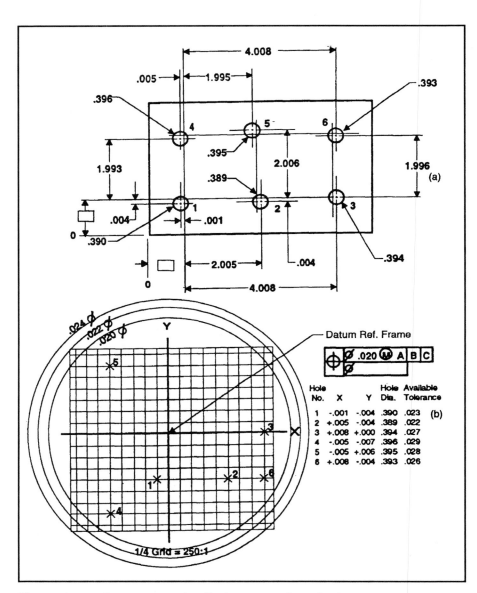

Figure 8-5.   Paper gage plot (hole pattern location).

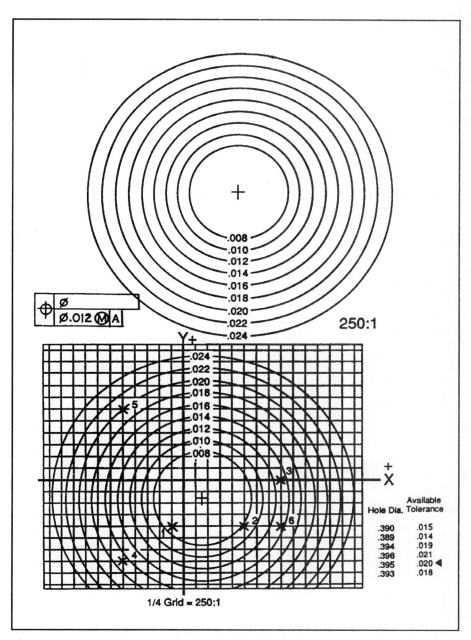

We are now ready to evaluate the lower control callout of hole-to-hole location tolerance. The tolerance is .012 diameter at MMC, relative only to datum A (perpendicularity). Therefore, the holes are free to float within the previously established .020 diameter tolerance zones. (See Figure 8-6.)

With our same-scale overlay, we can determine that the holes are acceptable for size and that the part is within tolerance.

Figure 8-6.   Paper gage plot (individual hole locations).

So, a typical gaging process could be:

1. Observe the plot of the hole pattern. We see hole #5 is displaced up and to the left from the rest of the holes in the pattern. We could consider hole #5 as a worst case for the purpose of placement of the overlay to save time.

2. The available hole tolerances have been determined and bonus tolerances applied. The overlay is placed with the .020 diameter tolerance tangent to the center of hole #5, insuring it is within tolerance. Notice this allows the overlay center to move down and to the right by .0015 (1½ squares).

3. This process accepts the same quality level as the functional gage. We have allowed for size and bonus tolerance and the hole-to-hole tolerance to be independent from the part edges. Because the tolerance zones are diameters, we have gained 57% area tolerance for manufacturing.

We are now ready to try an exercise in paper gaging, only in this case we will use a circular pattern of holes.

In Figure 8-7, the six holes are located relative to the datum framework B, A, and C, all RFS. Further, the holes are to be relative to one another and perpendicular to B within a .005 diameter at MMC.

To locate the part for measurement, we locate on primary surface B, center on secondary datum A (RFS), and stop rotation by use of datum C at RFS. (See Figure 8-8.)* To continue the process, we would

1. Plot hole locations, record hole size, determine available tolerance and determine if, up to this point, that part is acceptable, per the upper control.

2. From the lower control callout (see Figure 8-9), plot hole locations, note recorded hole size, determine available tolerance, and determine if the part is acceptable.

Let us review our example:

• Are the hole locations plotted correctly?
• Since the upper position control specifies the position control relative to datum A (RFS), our overlay must be concentric with the axis of the datum frame work RFS. The plot cannot be moved about, so the part is OK so far.

Now check the plot of holes for the lower callout. Only datum B is invoked. The plot of holes is free to float within the tolerance

*A transparent overlay to the same scale as the plot is required for this exercise.

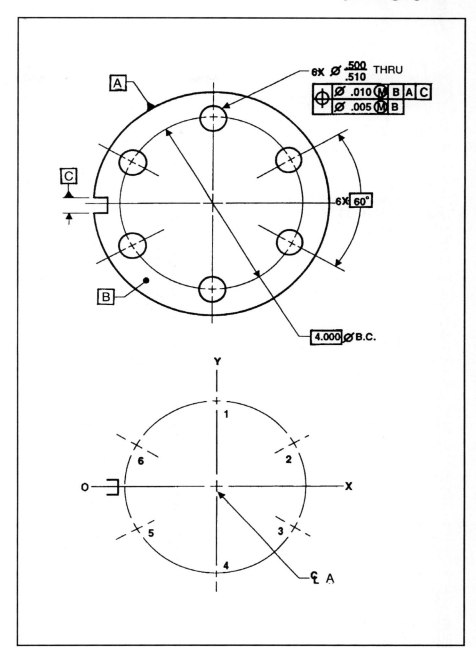

Figure 8-7.

of the upper callout, but hole-to-hole tolerance and perpendicularity must be within a .005 diameter at MMC. Since holes #1 and #6 are the farthest away from each other in the group, we can place the overlay on a line connecting holes #1 and #6. The best possible location for the overlay would be with holes #1 and #6 as close to the overlay center as possible (.0105).

Since the holes do not meet the requirements of the lower control, is the part scrap or can it be salvaged? The hole size is

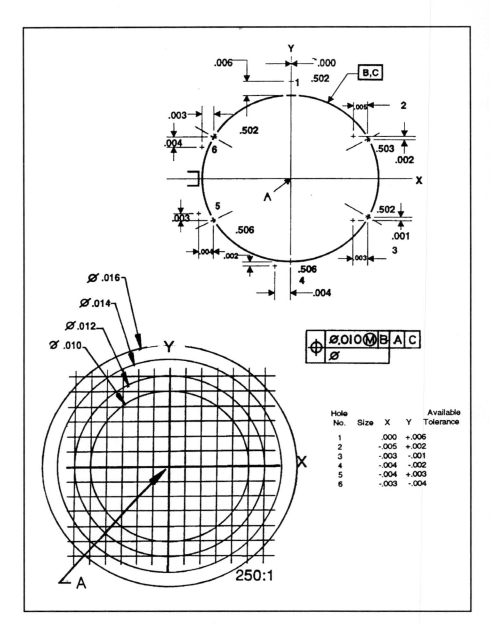

Figure 8-8.   Paper gage plot (hole pattern location).

.500 to .510 diameter or .510 at LMC. If we redrill the holes to a .506 diameter, what is the location tolerance (.011)? Would all holes then be within tolerance? The answer is "yes."

Take note that this method has not allowed for out of squareness in the holes to the surface B. We must consider this in order to accurately determine fit with the mating part. One method is to plot both the entry and exit of holes (if they are thru holes). Another technique is to consider the machine capability. If the fixture and equipment is capable of holding squareness within .002 per inch, and the mating part is .500 thick, we could add .001 to each hole plot for a worst-case comparison before using the over-

Template

250:1

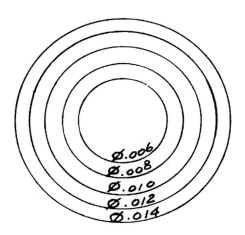

⌀.006
⌀.008
⌀.010
⌀.012
⌀.014

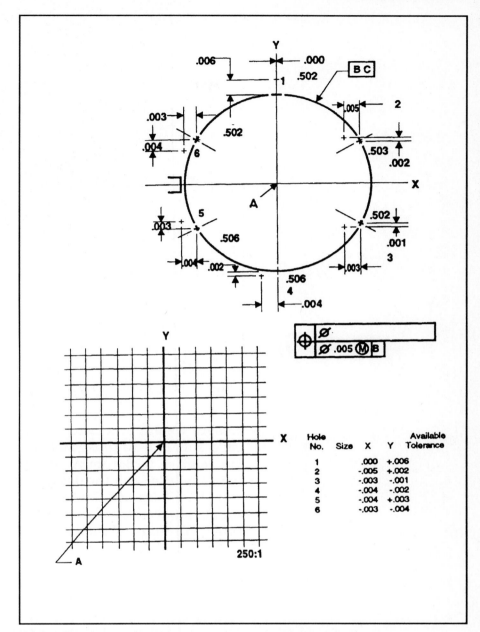

Figure 8-9.   Paper gage plot (individual hole locations).

lay. CMMs, along with appropriate software, make this job much easier. Formulas for determining squareness of features may be found on page 218 and in ASME Y14.5, Appendix B.

Open Set-Up Inspection Versus Functional Gage

Not all designs and parts have sufficient volumes to justify functional gages. However, when accepting or rejecting parts we must

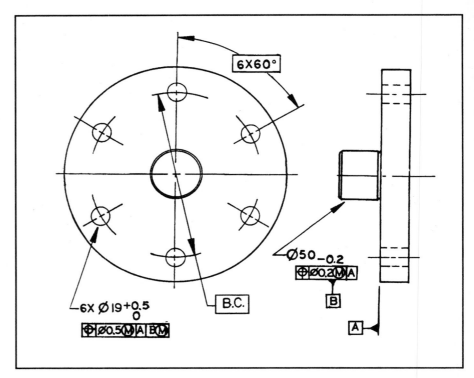

Figure 8-10.   Position tolerancing.

be sure we are evaluating to the same criteria as the functional gage. Figure 8-10 illustrates a part with six circular holes which are relative to datum A (surface) and datum B (center shaft). Both the holes and the datum have location tolerance. Tolerances and the effects of MMC for features are not difficult to understand, however, the application of datum tolerances and effects of bonus tolerance occasionally are misapplied.

Figure 8-11a shows the tolerance zones for features accepted by the functional gage on a part made to LMC conditions. There exists a 0.4 total tolerance at the axis of datum B at LMC, along with a tolerance of 1.0 at the true position axis of each hole. The axis of the hole pattern is free to move about the axis of the datum as influenced by the LMC conditions on both the feature holes and the datum. Figure 8-11b illustrates an incorrect application of the datum tolerance and bonus tolerance which would accept a greater tolerance level at the axis of each hole but with no translation or shift of the hole pattern relative to the datum. A differing level of accuracy may be accepted by this process as shown in Figure 8-12.

Figure 8-12a demonstrates the application of datums and tolerances that are comparable to the functional gage. Compare the results of both plots in Figure 8-12. Note that in the upper plot, the 1.0 feature zone cannot accept both holes 1 and 6 at the same time, however, the 1.0 tolerance zone is allowed to translate about the datum tolerance of 0.4 diameter. The lower plot has

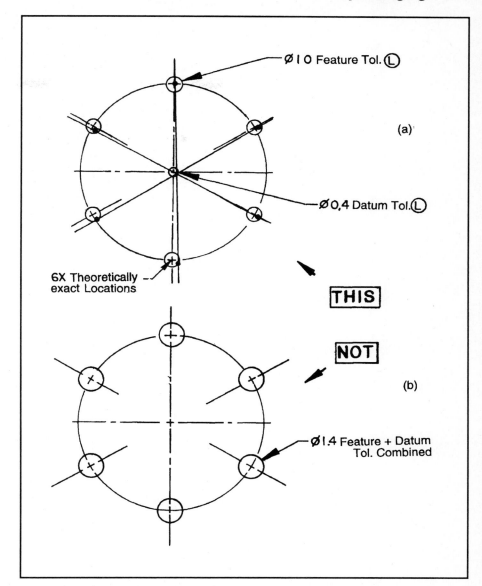

Figure 8-11.   Datum and feature tolerance zones at LMC.

combined the feature and datum tolerances giving an acceptance level of 1.4 diameter applied at each hole. Which method is similar to the gage? Figure 8-12a rejects the part, while Figure 8-12b accepts it. The part is out of spec. The method which accepts or rejects as the same quality level as the gage is correct.

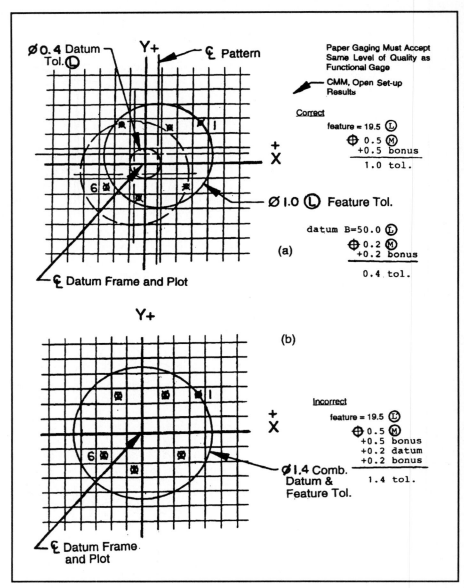

Figure 8-12.   Datums and feature tolerance zones—LMC (paper gage plot).

Paper Gaging Advantages

- Immediate analysis
- Low cost
- Independent of lot size (volume)
- No gage design required
- Provides process data
- No lead time
- No wear allowances
- No storage requirement
- Precise inspection report set-up

Summary

Let's review the key issues of paper gaging. As paper gaging is not dedicated or fixed, it may be applied to prototypes or very low volume parts, including audits. Performed properly, it will give us information from a controlled set-up, thus giving us information on process data, fixture accuracy and spindle wear. It also affords us the advantage of immediate analysis and feedback for process control or for supplier information. Gage design, gaging tolerances, and storage costs are also avoided.

Paper gaging is not the answer for everything or every situation. However, given proper thought, its use can compliment a quality control system, giving an extra tool to engineering, quality, and manufacturing.

9

Other Controls/Conventions

Dimension Origin Symbol

The dimension origin symbol was added to the 1982 standard to replace an arrowhead, when needed, on older drawings in order to clarify where the tolerance zone exists. Drawings made from former practices did not always have datum references. It was often difficult to determine if the tolerance existed at the foot or at the hole axis as shown in Figure 9-1. This symbol, when added to older drawings, must carry a note so the drawing user knows how to read and interpret the drawing requirements. Such a note might read: 1. See ASME Y14.5M-1994. The intention of this symbol is not to replace the datum system, so it is permissible to use this symbol on newly created drawings for clarity.

Hole Depth Dimensioning (Curved Surfaces)

Hole depths on curved surfaces are often difficult to evaluate with conventional gaging methods. When holes are produced in curved surfaces, the technique shown in Figure 9-2 is recommended as a dimensioning option. This technique would allow a gage pin of given length to be inserted into the hole. The measurement is taken over the pin length from the opposite side of the shaft. The pin length is subtracted from the measured length to determine hole depth. This is the same dimensioning principle used on key-slots and key-ways.

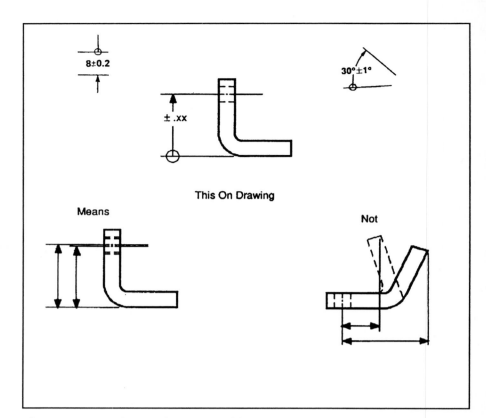

Figure 9-1.   Relating dimensional limits to an origin.

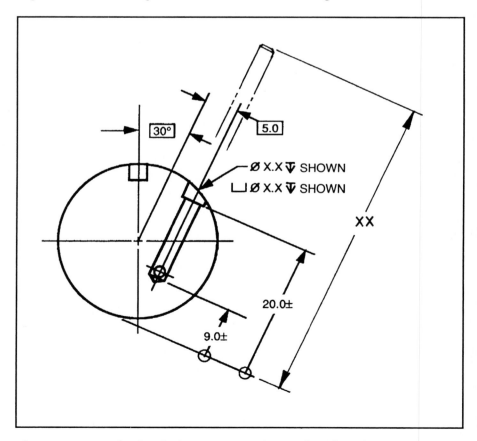

Figure 9-2.   Hole depth dimensioning (curved surfaces).

Projected Tolerance Zone

Projected tolerance zones "project" the normal position or perpendicularity tolerance zone above (or below) the part as indicated by the call-out or other necessary detail. The application of a projected tolerance zone is recommended where the variation in the perpendicularity of threaded or press fit holes could cause fasteners to interfere with mating parts. An interference can occur where a threaded or press fit fastener is inclined within the position tolerance limits as shown in Figure 9-3. Unlike the floating fastener application involving clearance holes only, the attitude of a fixed fastener is governed by the inclination of the produced hole into which it assembles. Optional methods of application of the projected zone symbol and feature control callout are shown in Figure 9-3.

When the projected tolerance system is not used, it will be necessary to select a position tolerance and hole size that insures a clearance fit. The formula for allowing for this out of squareness is:

$$H = F + T_1 + T_2 \left(1 + \frac{2P}{D}\right)$$

where
T_1 = position tolerance diameter of clearance hole
T_2 = positional tolerance diameter of threaded or press fit hole
D = depth of threaded or press fastener
P = maximum fastener projected height
H = clearance hole size
P = fastener diameter

Taper and Slope

Taper. A *taper* is defined as the ratio of the difference in the diameters of two sections (perpendicular to the axis) of a cone to the distance between these sections.

Slope. A *slope* is specified as the inclination of a surface expressed as a ratio of the difference in the heights at each end (above and at right angles to a base line) to the distance between those heights.

Conical tapers. These include standard machine tapers used throughout the tooling industry which are classified as American Standard Self-Holding and Steep Taper series. (See ANSI B5.10.)

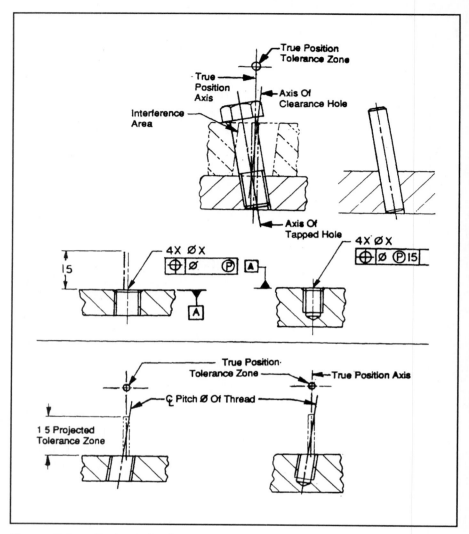

Figure 9-3. Projected tolerance zone.

The *self-holding series* has 22 sizes: three small sizes with a taper of ½ inch per foot; eight sizes with a taper of ⅝ inch per foot; and 11 sizes with a taper of ¾ inch per foot. The *steep taper series* has 12 sizes and are defined as having a sufficiently large angle to insure an easy to self-releasing feature.

American Standard Machine Tapers are generally dimensioned using the taper name and number, with reference to the taper per foot. The diameter at the gage line may be given, as well as length, with the small end of the taper as a reference. (See Figure 9-4.)

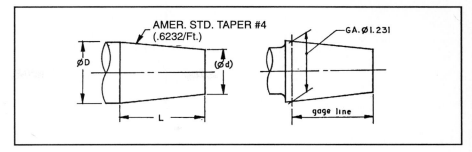

Figure 9-4. American standard machine taper.

A taper may also be illustrated by the method used in Figure 9-5. For more complete information on machine tapers, refer to ANSI B5.10.

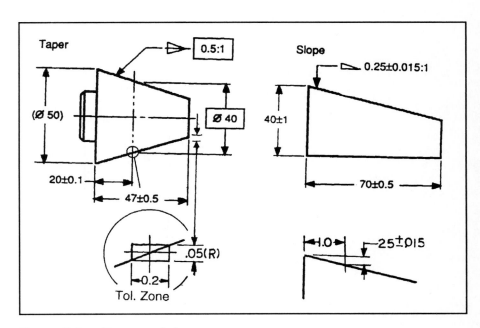

Figure 9-5. Taper and slope.

Spherical Features

Spherical features are features of size. They are continued surfaces and may have size, form, and/or location error. They have size tolerance and they may require some refinement of size, such as profile. In addition, the feature may require a tolerance of location (MMC or RFS). Figure 9-6a illustrates a spherical feature with a size tolerance of 40.0 to 39.5 diameter and a refinement of a profile tolerance, within size limits, of 0.2 between points X-Y. The profile tolerance is for the full feature surface

(surface profile symbol) and must fall within the size tolerance. The spherical feature must have a location tolerance of 0.4 spherical diameter relative to primary datum feature B (RFS) and secondary datum surface A. When the diameter symbol in the feature control frame is related to a spherical feature, the tolerance zone is spherical.

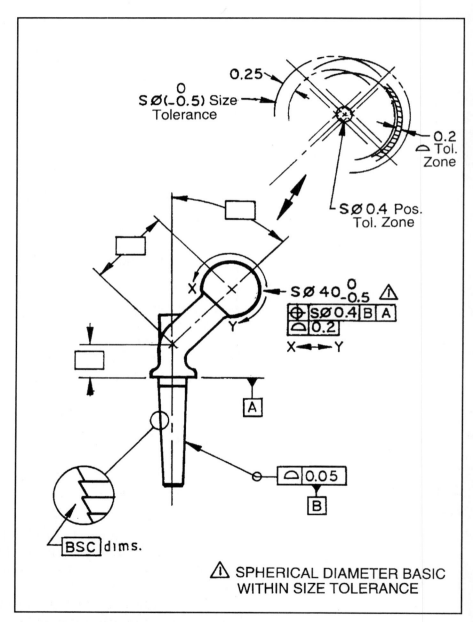

Figure 9-6a.   Spherical features.

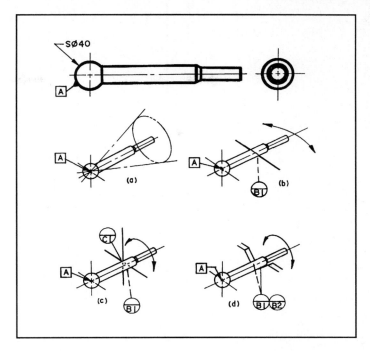

Figure 9-6b.

If spherical features are chosen as primary datum features, it will become more difficult to stabilize the part. Figure 9-6b illustrates a shaft with a spherical ball end as the primary datum feature A. With only datum A invoked, the part is free to rotate freely in space, anchored only by A. If line contacts were made on the shaft diameter, one degree of freedom is removed, however, the part is still free to rotate from side to side. By adding a second line contact (90° to B1), we have stopped the rotational movement side to side. Note the axial rotation is still possible. This is because the primary datum is only a point. In these cases, if it is determined the primary datum feature must be a spherical feature (datum point), a tertiary datum keyslot or a flat on the shaft could be used to stop axial rotation.

Other Symbols

Controlled radius, or CR, is a new concept for the 1994 standard which allows a variable radius, but without flats or reversals as shown in Figure 9-7. A *controlled radius* creates a tolerance zone defined by two arcs (the minimum and maximum radii) that are tangent to the adjacent surfaces. When specifying a *controlled radius*, the part contour within the crescent-shaped tolerance zone must be a faired curve without reversals. Note: In previous standards the symbol R implied no flats or reversals. Other symbols are also shown in Figure 9-7 which will be often used on drawings. Note the arrowhead symbols from ANSI/ASME Y14.2 on line conventions and lettering.

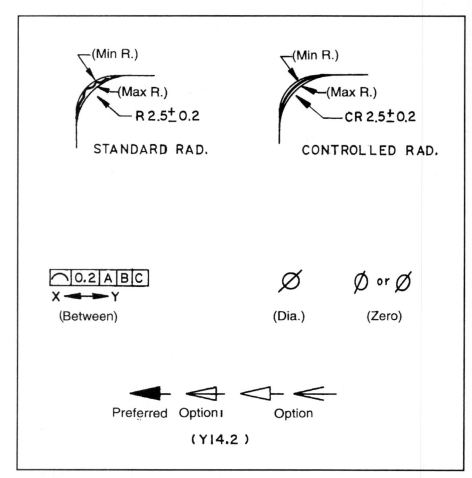

Figure 9-7.   Controlled radius and other symbols.

Symmetrical Outlines

Figure 9-8 illustrates the convention for dimensioning one side of a centerplane of symmetry. This convention is also covered by ANSI Y14.2 and is shown here for reference.

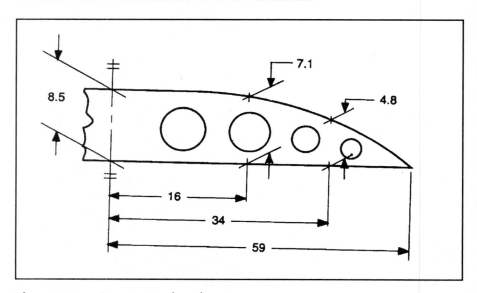

Figure 9-8.   Symmetrical outlines.

Chamfers

While doing a study on chamfers, it was determined there are four elements to a chamfered feature (shaft or hole), three of which may have a tolerance. The remaining element would become a reference or a result of the other dimensions as shown in Figures 9-9. The hole (or shaft) diameter is always toleranced, as it is a feature of size, and the basis for all other dimensions. The remaining dimensions may be toleranced or basic, depending on the design requirements, with the resulting chamfer diameter B generally being a result of the other dimensions and, therefore, a reference.

My manufacturing friends have told me that for most designs, they would prefer to see angle C ± and the depth D as basic for internal chamfers. For external chamfers, they recommend a basic angle C with depth ±. The standardized recommendations were based on assumptions they had no detrimental effect on design integrity and are included here for reference information only.

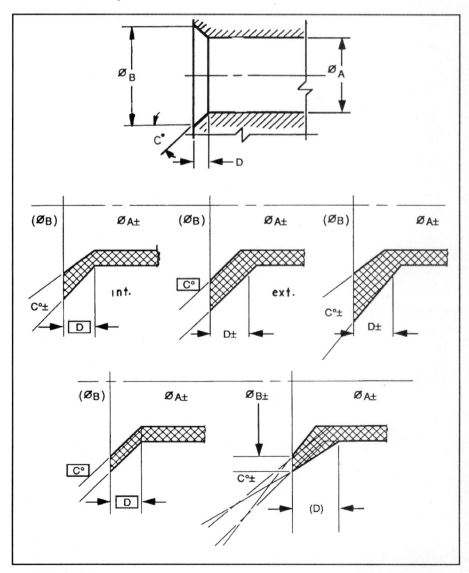

Figure 9-9.   Chamfers.

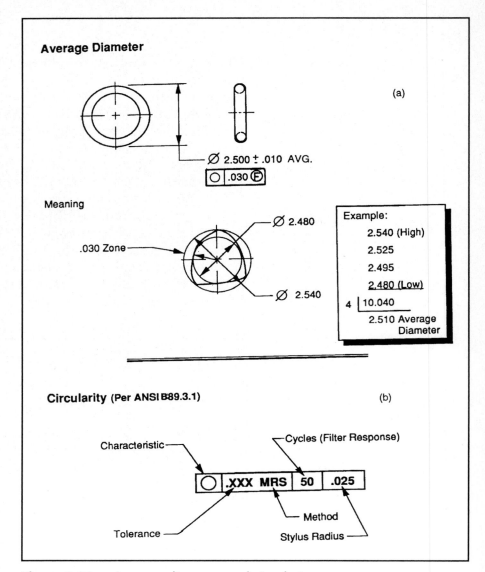

Figure 9-10. Average diameter and circularity.

Average Diameter

Specifying *average diameter* and roundness on a non-rigid part may be necessary to insure that the actual diameter of the feature can be controlled to the desired shape with the parent parts (mating parts) at assembly. The average diameter is the average of several diametrical measurements across the feature in its free or unrestrained state. Where practical, the average diameter may be determined by peripheral measurement. The free state roundness control value will be larger than the average diameter tolerance when used in this combination. Figure 9-10a illustrates an acceptable part with these controls specified. The circle F (Ⓕ) symbol indicates the acceptable roundness limit in the free state condition.

Circularity Critical Features

Circularity has a separate standard regarding roundness measurement techniques and inspection methods. ANSI B89.3.1 contains specific information on roundness measurement when specified on a drawing as shown in Figure 9-10b. The feature control frame consists of the roundness symbol, tolerance, assessment method (there are four), cycles/filter response, and stylus tip radius. This control is not common to most part requirements. I recommend that you review the B89.3.1 standard before applying the concepts. This control is used for critical components where strict conformance and repeatability are a must and/or where measurement techniques are controlled. Consult your Quality Control Department for further assistance.

Free State Variation

Free state variation is a term used to describe distortion of a part after removal of forces applied during manufacture. This distortion principally is due to weight and flexibility of the part and the release of internal stresses resulting from fabrication. A part of this kind, for example, a part with a very thin wall in proportion to its diameter, is referred to as a *non-rigid part*.

Figure 9-11 illustrates the use of combined average diameter and free state roundness control for a datum along with a note ⚠1 for restraining the datum features while evaluating a related feature. This technique is often required on parts defined as "non-rigid." (See Figure 9-11.) Remember, specifications exist with datums and features in their free state unless otherwise noted.

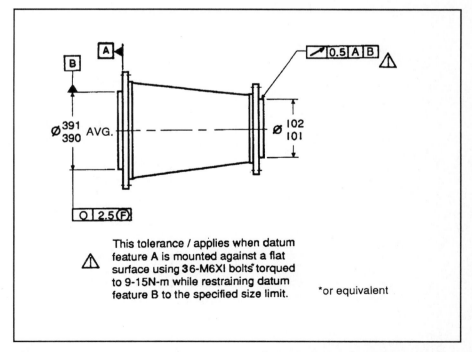

Figure 9-11.   Free state variation—specifying restraint for non-rigid parts.

Statistical Tolerance

Statistical tolerancing has generally not been used on product drawings, as it is considered in the realm of "process engineering" or "process information." The ASME Y14.5 standard has included the symbols and drawing techniques as a *recommendation for standardized application* of the principles should you choose to apply statistical tolerancing techniques to your engineering drawings. My personal recommendation is to use statistical tolerancing:

1. Only after exhausting direct tolerancing of 100% fit methods
2. As a joint design/manufacturing/quality effort
3. When multi-plant/supplier issues are minimal
4. Only when statistical process control documentation is in place.

Statistical tolerancing is the assigning of tolerances to related components of an assembly on the basis that the assembly tolerance is equated to the square root of the sum of the squares of the individual tolerances. When tolerances assigned by arithmetic stacking are restrictive, statistical tolerancing may be used for increased individual feature tolerance. The increased tolerance may reduce manufacturing cost, but should only be employed where the appropriate process control will be used. (See Figure 9-12a.)

It may be necessary to designate both the statistical limits and the arithmetic *stacking limits* when the dimension has the possibility of being produced without statistical process control (SPC). (See Figure 9-12b.)

The statistical tolerancing principles may also be applied to geometric tolerance controls as shown in Figure 9-13. In this example, statistical tolerancing techniques have been applied to

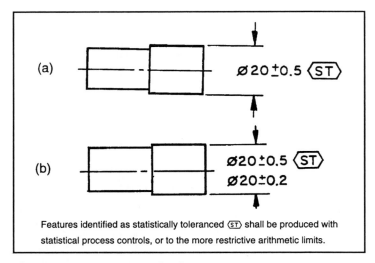

Figure 9-12.   Statistical tolerancing.

both the size and position controls. Generally, statistical tolerancing should be a colaberative process, with tolerances arrived at through the involvement and input of manufacturing and quality. Process control systems may vary slightly between companies or agencies, but, when shown as part of the drawing specifications, they become part of the contract between supplier and customer (or departments or agencies) and should be documented in a process control agreement or standard. It may also be a good idea to include, on the drawing, the note shown in Figure 9-12.

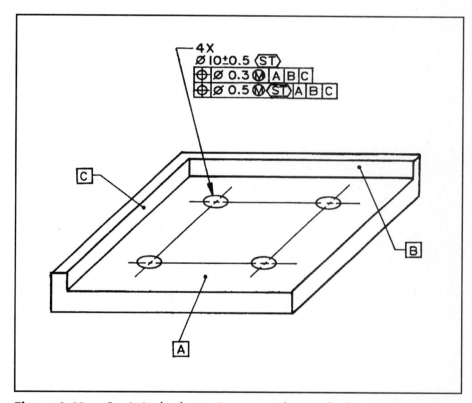

Figure 9-13.   Statistical tolerancing principles applied to geometric tolerance controls.

Virtual Resultant Conditions

Virtual and resultant conditions were discussed in an earlier chapter, and are defined again in the definition section and glossary. However, let's review the principles with an example.

Remember, MMC and LMC are opposite, holes and shafts are opposite, inner and outer boundaries are opposite, virtual condition occurs at MMC or LMC, and resultant condition contains the additive effects of all bonus size and geometric tolerances. The examples in Figure 9-14 may be either holes or shafts, at MMC or LMC, with both location bonus due to size applied. MMC gives the closest fit with mating components, while LMC will use up the most material.

Review Figure 9-14 for application to both holes and shafts, noting the effects of MMC and LMC in order to produce the potential inner and outer limit boundaries.

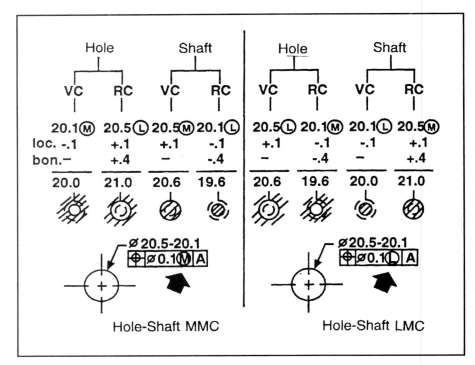

Figure 9-14.   Virtual/resultant conditions.

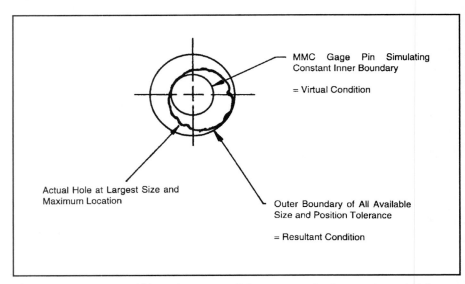

Figure 9-15.   Virtual/resultant condition as applied to an internal feature specified on an MMC basis.

Surface Texture

Although surface texture is not part of the Y14.5 standard, the specified feature tolerances on engineering drawings have a direct relationship to surface texture and visa versa. This relationship is due to specified tolerances and equipment capability. Figure 9-16 illustrates this relationship. Generally, it would not appear practical to specify a tolerance (size) of .010 with a surface finish of 8 µin, or a tolerance of .0004 with a surface finish of 250 µin. There may be exceptions, but equipment capability will have an impact on tolerancing versus surface finish.

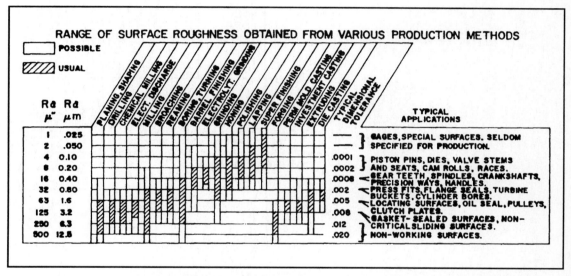

Figure 9-16. Range of surface roughness obtained from various production methods (see ANSI B46.1 and ANSI Y14.36.).

Information on surface texture, primary and secondary, may be found in ANSI B46.1, with the symbology covered in Y14.36. A comparison of micro inch, micro meter, ISO roughness grades, and former practices appears in Figure 9-17.

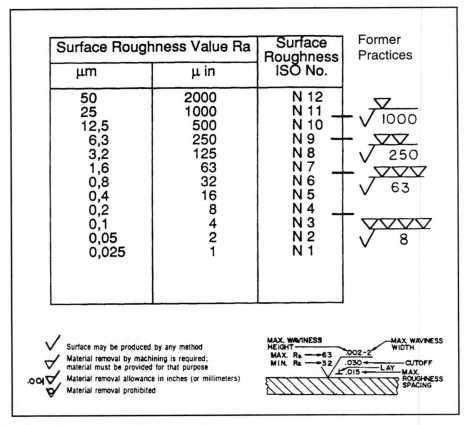

Figure 9-17.   Surface roughness values.

Appendix

Feature Control Selection Diagram

Figure A-1 is intended as an aid for engineers, designers, drafters, or others in determining which geometric controls would be most appropriate, based upon the industrial hierarchy technique of yes-no type questions and responses. Follow the design issue or feature in question on the chart until the requirement is resolved. This chart may not satisfy every circumstance, but with reasonable care most design tolerance issues should be covered. For complex designs, it will be necessary to go through the chart many times, for various features, until you are comfortable with the process.

Quality Control

The quality function is primarily geared toward the activities of acceptance of product and collecting variables data for process control purposes. As products are developed, so are quality plans for these products. The two key elements which help form these plans are the plant environment/personality and available measurement equipment or tools. A typical menu of issues is shown in Figure A-2 but is not all inclusive. Our quality plan will be influenced by these factors, which must be dealt with, hopefully as a joint effort by engineering, manufacturing and quality.

I have charted these issues much like the design diagram earlier. (See Figure A-3.) The order of issues may differ in importance relative to your organization or product, however, I believe similar issues will arise which must be resolved. The ideal situation would be where design change is minimal, where we have skilled people in critical positions, and where we have the full range of gaging options at our disposal. Work through the chart as before. Are there other issues you wish to include?

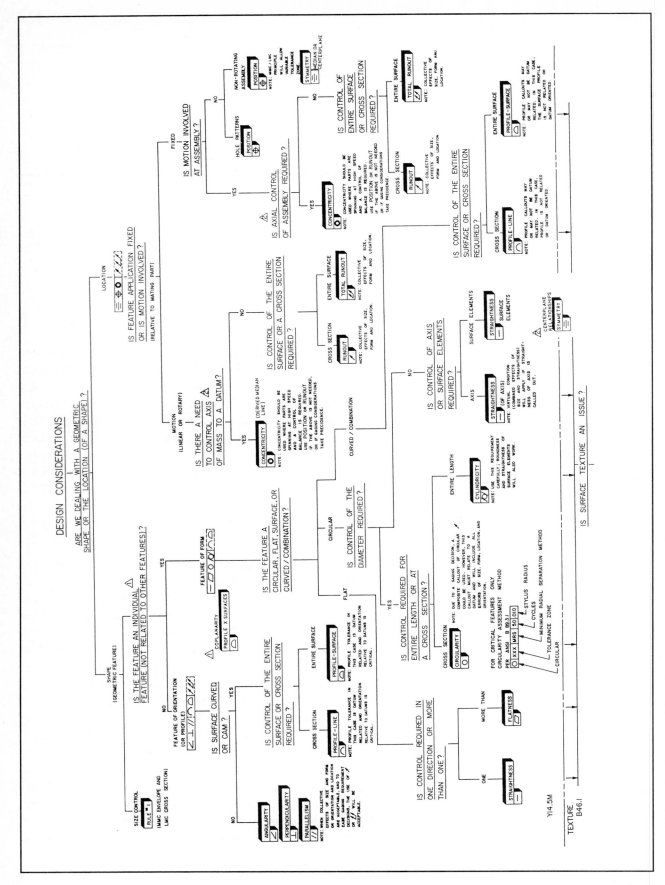

Figure A-1. Design considerations.

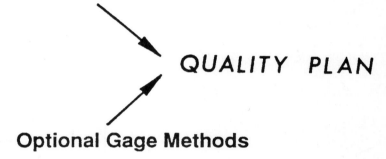

Plant Considerations

- Needs/Requirements

- Volumns

- Risk Exposure

- Warranty Record

- Equipment Maintenance Schedule

- Personnel Systems

- Others

QUALITY PLAN

Optional Gage Methods

- CCM Inspection W/WO Programmable Software

- Surface Plate Inspection

- Functional Gaging

- Mathematical Analysis

- Equipment Audit-No Part Inspection

- Combinations of All

Figure A-2.   Quality plan diagram.

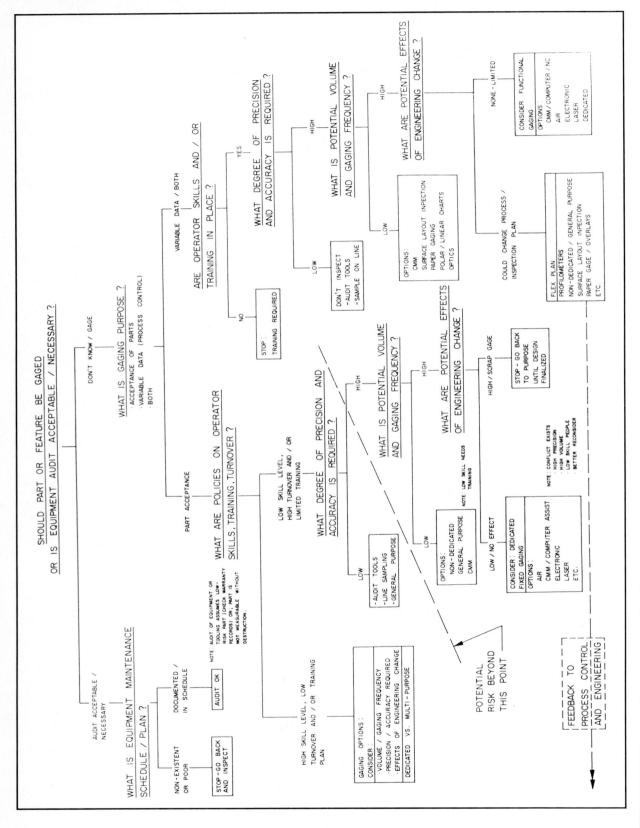

Figure A-3.   Quality plan considerations.

ANSI B4.1	Preferred Inch Limits and Fits
ANSI B4.2	Preferred Metric Limits and Fits
ASME Y14.1M	Drawing Sheet Size and Format
ASME Y14.2	Line Conventions and Lettering
ASME Y14.3M	Multi-Sectional View Drawings
ASME Y14.4M	Pictorial Drawings
ASME Y14.5M	Dimensioning and Tolerancing
ASME Y14.5.1M	Mathmatical Definition (Y14.5M)
ANSI Y14.6	Screw Thread Representation
ANSI Y14.7	Gears and Splines
ASME Y14.8	Castings and Forgings
ANSI Y14.11	Molded Parts Drawings
ANSI Y14.15	Electronic and Electrical Diagrams
ANSI Y14.17	Fluid Power Diagrams
ANSI Y14.18M	Optical Parts Drawings
ASME Y14.24M	Types and Application—Engineering Drawings
ANSI Y14.31	Undimensioned Drawings
ANSI Y14.36	Surface Texture Symbols
ASME Y14.37	Composite Parts Drawings (prop)
ASME Y14.100	MIL-STD 100 (prop)
ANSI Y1.1	Abbreviations
ANSI/IEEE 268	Metric Practice

Note: ASME/ANSI is in process of re-publishing many of the above standards. When a new standard is issued, it will carry the ASME prefix.

Figure A-4.   American National Standards References.

ASME B1.2	Gaging for "UN" Screw Threads (pin gages)
ANSI B1.3	Gaging Systems for Screw Thread Acceptability
ANSI B1.16	Gaging Practices for Metric Screw Threads
ANSI B1.20.5	Gaging for Pipe Threads
ANSI B1.22	Gaging Practices for "MJ" Series Metric Threads
ANSI B5.10	Standard Machine Tapers
ANSI B46.1	Surface Texture
ASME B47.1	Gage Blanks
ANSI B89.1.6	Master Rings and Ring Gages
ANSI B89.1.9	Precision Gage Blocks Through 20 in.
ANSI B89.6.2	Temperature and Humidity Environment for Measurement
ASME B92.1	Involute Splines and Inspection
ASME B94.6	Knurling
ANSI B94.11M	Drills
ANSI B4.4	Inspection of Workpieces
ANSI B89.3.1	Roundness Measurement
ANSI B89	Parametric Calibration of CMM's
ANSI B89.3.6	Functional Gaging (New)

Figure A-5.   American National Standards related to dimensional metrology.

Quantity	Multiply	By or Use Formula	To Obtain Equivalent Number of:	
acceleration	ft/sec^2 in/sec^2	0.304 8 0.025 4	**m/s^2** **m/s^2**	(metre/second squared)
area	ft^2 in^2	0.092 9 6.452×10^{-4}	**m^2** **m^2**	(square metres)
bending moment or torque	kg$_f$–m oz$_f$–in lb$_f$–in lb$_f$–ft	9.807 7.062×10^{-3} 0.113 1.356	**N•m** **N•m** **N•m** **N•m**	(newton • metre)
density	lb$_m$/ft^3 lb$_m$/in^3 lb$_m$/gal (U.S.) lb$_m$/gal (Imp.)	16.03 27.70×10^3 0.119 9 0.099 9	**kg/m^3** **kg/m^3** **kg/l** **kg/l**	(kilogram/cubic metre)
energy, work and heat	btu lb$_f$–ft	1.055 1.356	**J** **J**	(joule)
flow meter	cts/gal (U.S.) cts/gal (Imp.) cts/ft^3	0.264 0.220 3.531×10^{-2}	**cts/l** **cts/l** **cts/l**	(counts/litre)
flow rate	lb$_m$/hr lb$_m$/hr *gm/hr *gm/min	1.26×10^{-4} 0.454 2.778×10^{-7} 1.667×10^{-5}	kg/s **kg/hr** kg/s kg/s	(kilogram/second) (kilogram/hour)
			*gm/hr and gm/min are preferred units requiring no conversion.	
force	kg$_f$ kilopond$_f$ lb$_f$ poundal$_f$	9.807 9.807 4.448 0.138 3	**N** **N** **N** **N**	(newton)
			up to 1000 lb$_f$ use N (newton) 4.448 from 1000–100,000 lb$_f$ use kN (kilonewton) 4.448×10^{-3} above 100,000 lb$_f$ use MN (meganewton) 4.448×10^{-6}	
fuel performance	miles/gal (U.S.) miles/gal (Imp.) miles/qt (U.S.) miles/qt (Imp.) gal/mile (U.S.) gal/mile (Imp.) lb$_m$/bhp•hr lb$_m$/bhp•hr	0.425 1 0.353 9 1.701 1.415 6 2.352 2.825 0.608×10^{-3} 0.608	**km/l** **km/l** **km/l** **km/l** **l/km** **l/km** kg/W•hr **kg/kW•hr**	(kilometre/litre) (litre/kilometre) (kilogram/watt•hour) (kilogram/kilowatt•hour)
length	ft in mile	0.304 8 25.4 1.609	**m** **mm** **km**	(metre) (millimetre)—all engineering drawings will be dimensioned in mm (kilometre)
mass	oz (avoir) lb (avoir) slug	0.028 4 0.454 14.6	**kg** **kg** **kg**	(kilogram)
moment of inertia (2nd moment of area)	lb$_m$–ft^2 lb$_m$–in^2	4.217×10^{-2} 2.929×10^{-4}	**kg•m^2** **kg•m^2**	(kilogram•metre squared)
power and heat rejection	hp btu/min	746 17.58	**W** **W**	(watt)
			below 1 hp use W (watt) 746 above 1 hp use KW (kilowatt) 0.746	

Figure A-6.   Metric conversion table.

Quantity	Multiply	By or Use Formula	To Obtain Equivalent Number of:		
pressure	in Hg	25.4	**mm Hg**	(millimetre of mercury)	
	in Hg @ 32°F	3.386×10^3	Pa	(pascal)	
	in Hg @ 60°F	3.377×10^3	Pa		
	in H_2O	25.4	**mm H_2O**	(millimetre of water)	
	in H_2O @ 39.2°F	249	Pa		
	in H_2O @ 60°F	248.84	Pa		
stress	kg_f/mm^2	9.8×10^6	**Pa**		
	lb_f/ft^2	47.88	**Pa**		
	lb_f/in^2(psi)	6.894×10^3	***Pa**		
	poundal/ft^2	1.488	**Pa**		
			*up to 1 psi, use Pa (pascal) 6.894 $\times 10^3$		
			from 1—1000 psi, use kPa (kilopescal) 6.894		
			above 1000 psi, use MPa (megapascal) 6.894×10^{-3}		
temperature	**°Celsius**	$t_{o_C} + 273.15$	K	(kelvin)	
	°Fahrenheit	$(t_{o_F} + 459.67)/1.8$	K		
	°Rankine	$t_R/1.8$	K		
	°Fahrenheit	$(t_{o_F} - 32)/1.8$	**°C**	**(Celsius) (Centigrade)**	
	°Kelvin	$t_K - 273.15$	°C		
temperature interval	**°Celsius**	1.0	K		
	°Fahrenheit	5.556×10^{-1}	K or °C		
velocity	ft/min	5.08×10^{-3}	**m/s**	(metre/second)	
	ft/sec	0.304 8	**m/s**		
	km/hr	$0.277\ 8 \times 10^{-7}$	**m/s**		
	mile/hr	1.609	**km/hr**	(kilometre/hour)	
viscosity	Centipoise	0.001	**Pa•s**	(pascal•second)	
	Centistokes	1.00×10^{-6}	**m^2/s**	(square metre/second)	
volume	fl oz (U.S.)	2.957×10^{-2}	l	(litre)	
	quart	0.946	l		
	gal (U.S.)	3.785	l		
	gal (Imp.)	4.546	l		
displacement	in^3	1.639×10^{-2}	*l		
	**in^3	1.639×10^{-5}	**m^3**		
	**ft^3	2.832×10^{-2}	**m^3**		
	**ft^3	28.32	l		
	litre	0.001	m^3		
	m^3	1 000	l		
			*litre is the preferred unit for engine displacement		
			**solid displacement		
volumetric flow	ft^3/min	28.32	*l/min	(litre/minute)	
	ft^3/min	0.472	*l/s	(litre/second)	
	ft^3/sec	28.32	l/s		
	in^3/min	2.73×10^{-4}	l/s		
	oz/hr	2.957×10^{-2}	l/hr		
	qt/hr (U.S.)	0.946	l/hr		
	qt/hr (Imp.)	1.1365	l/hr		
	gal (U.S.)/min	3.785	l/min		
	gal (U.S.)/min	0.063 1	l/s		
	gal (U.S.)/hr	3.785	l/hr		
	gal (Imp.)/min	4.546	l/min		
	gal (Imp.)/min	0.075 8	l/s		
	gal (Imp.)/hr	4.546	l/hr		
			*from 1–100 ft^3/min, use l/min		
			above 100 ft^3/min, use l/s		
weight to power	lb/hp	0.608×10^{-3}	kg/W	(kilogram/watt)	
	lb/hp	0.608	**kg/kW**	(kilogram/kilowatt)	

Note: Boldface type indicates preferred units.

Figure A-6. *(cont.).*

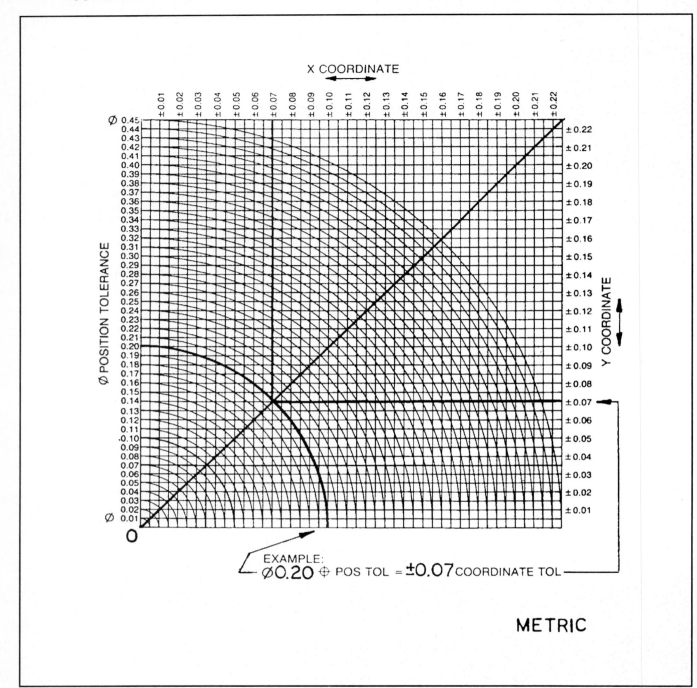

Figure A-7. Conversion chart—position tolerance to coordinate, coordinate tolerance to position.

Resource Information

ANSI
1430 Broadway
New York, NY 10018
212-354-3300

SME
One Dearborn Drive
P.O. Box 930
Dearborn, MI 48121-0930

National Center for Standards
 and Certification Information
National Institute for Standards
   and Technology
U.S. Department of Commerce
Administration Building, Room A629
Gaithersburg, MD 20899
301-975-4040

International Standards Organization (ISO)
Rue de Varembe 1
CH-1211 Geneva 20, Switzerland
(41 22) 34 12 40

American Society for Quality Control
310 W. Wisconsin Avenue
Suite 500
Milwaukee, WI 53203
414-272-8575 or 1-800-248-1946

ASME Order Department
22 Law Drive
Box 2900
Fairfield, NJ 07007-2900
1-800-843-2763

Gary Gooldy
3240 Hillcrest Drive
Columbus, IN 47203
812-372-9693

ISO REFERENCE STANDARDS

ISO 1   STANDARD REF. TEMPERATURE FOR LENGTH
                 MEASUREMENT.

ISO 286-1   ISO SYSTEM OF LIMITS & FITS-PART 1.BASIS

ISO 286-2   ISO SYSTEM OF LIMITS & FITS-PART 2.TABLES

ISO 1938-1   ISO INSPECTION OF WORKPIECES-LIMIT GAUGES

ISO 1938-2   ISO INSPECTION OF WORKPIECES

ISO 1938-3   ISO INSPECTION OF WORKPIECES-GUIDELINES FOR INSPECTION.

ISO 3274   ISO INSTRUMENTS FOR PROFILE MEASUREMENT.

ISO 3650   ISO GAUGE BLOCKS

ISO 3670   ISO BLANKS FOR PLUG & RING GAUGES-GENERAL

IISO 8015   ISO FUNDAMENTAL TOLERANCING PRINCIPLES

ISO 1101   ISO TECHNICAL DRAWINGS-GEOMETRICAL TOLERANCING

ISO 128   ISO TECHNICAL DRAWINGS-GENERAL PRINCIPLES

ISO 129   ISO TECHNICAL DRAWINGS-DEFINITIONS & SPECIAL INDICATIONS.

ISO 1660   ISO DIMENSIONING & TOL.-PROFILES

ISO 2692   ISO DIMENSIONING & TOL. .MMC PRINCIPLE

ISO 5458   ISO DIMENSIONING & TOL. .POSITION TOL.

ISO 5459   ISO DIMENSIONING & TOL. .DATUMS & DATUM SYST.

ISO/TR 5460   TOL. OF FORM, ORIENTATION, LOCATION & RUNOUT VERIFICATION
                 PRINCIPLES.

ISO 7083   ISO SYMBOLS FOR GD&T-PROPORTION & DIMENSIONS

ISO IS A WORLDWIDE FEDERATION OF ISO MEMBER BODIES. THE WORK OF INTERNA-
TIONAL STANDARDS PREPERATION IS CARRIED OUT BY TECHNICAL COMMITTEES.
DRAFT STANDARDS ADOPTED BY THE TECHNICAL COMMITTEES REQUIRE APPROVAL
BY AT LEAST 75% OF THE MEMBER BODIES VOTING.

General Tests

I conclude this workbook with a series of exercises. I would be interested in your comments about the book. Please contact me at the following address. Good luck in your efforts.

Gary Gooldy
GPG Consultants Inc.
3240 Hillcrest Drive
Columbus, IN 47203

GENERAL TEST 1

2X ∅.255-.265 (holes)

| ⊕ | ∅.005 Ⓜ | C | B | A |

1. The Maximum Material Condition of size of hole is:
   A. .250   B. .255   C. .260   D. .265
2. Where the hole size is ∅.264, the size of the tolerance zone is:
   A. ∅.005   B. ∅.010   C. ∅.012   D.∅.014
3. The virtual condition of the hole is:
   A. ∅.250   B. ∅.255   C. ∅.260   D. ∅.265
4. The shape of the tolerance zone is:
   A. a width   B. a cylinder   C. a square.
5. The primary datum reference in the feature control frame is:
   A. datum feature A   B. datum feature B   C. datum feature C.
6. The names of the following symbols are:
   A. ⊕ _____   B. ⊥ _____   C. Ⓛ _____
   D. ∅ _____   E. ⌒ _____   F. ↗ _____
7. In a feature control frame, geometric controls are applied (at MMC) (at LMC) (at RFS) _____ unless otherwise specified.
8. How many minimum points of contact are needed to establish datum planes?
   A. Primary _____   B. Secondary _____
   C. Tertiary _____
9. A basic dimension:
   A. establishes a theoretically perfect feature size of location
   B. directs you to the tolerance block.
   C. establishes a feature MMC.
10. Concentricity is best used as a control for:
    A. position at MMC   B. coaxiality   C. interchangeability
11. Concentricity and Runout controls are to apply:
    A. at LMC   B. at MMC   C. regardless of feature size
    D. any of above
12. RFS applied to a tolerance value means that:
    A. The tolerance zones get smaller as the features get smaller.
    B. The tolerance zones remain the same size at any increment of feature size.
    C. The tolerance zones increase as the feature depart from their MMC size.
13. Tolerance zone values in a feature control frame are understood to be:
    A. plus and minus   B. totals   C. a bonus

14. Datums are:
    A. actual part surfaces   B. alphabetical features
    C. theoretically exact points axes and planes.

15. Perpendicularity, Angularity, and Parallelism controls govern the:
    A. shape   B. location   C. orientation between features.

16. Tolerance zones are:
    A. sometimes   B. always
    C. never located and oriented relative to datums.

17. If the primary datum is an axis and the secondary datum is a plane, the secondary datum feature may contact the datum reference at
    A. one-point   B. two-points   C. three-points

18. Where a datum feature is a plane surface, measurements from that feature to other features of the part are taken from:
    A. the actual surface of the part
    B. the datum simulated from that datum feature
    C. the theoretical datum surface.

19. The letters LMC stand for:
    A. Lease Manufacturing Cost
    B. Least Machining Consideration
    C. Least Material Condition

20. RULE #1...
    A. controls one feature relative to another.
    B. controls MMC, LMC and RFS.
    C. controls perfect form at MMC.
    D. controls the perpendicularity between plane surfaces.

21. From the drawing below, find minimum distance **"X"**

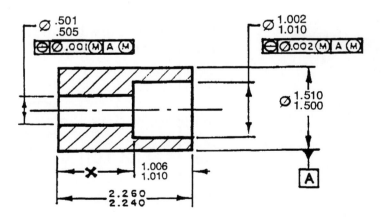

FROM THE DRAWING ABOVE:

22. Using the small hole on the left, what size would the POSITIONAL tolerance zone be if the actual hole size is .503?____

23. Using the hole on the right, what size would the POSITIONAL tolerance zone be if the actual hole size is 1.009?____

GENERAL TEST 2

Indicate below the pin to be perpendicular to the bottom surface within 0.25 ⌀ MMC.

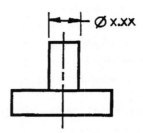

Complete the table below for the measured sizes given. (Axial straightness tolerance zone)

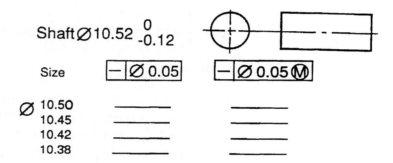

Complete the table below for the measured sizes given. (Perpendicular tolerance zone)

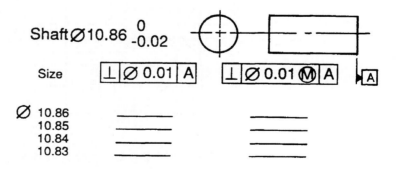

GENERAL TEST 3

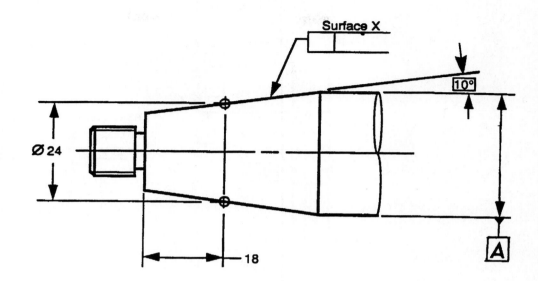

Required : (at RFS)

The <u>surface X</u> must lie between (2) coaxial boundaries 0.05 apart, having an angle of 10° relative to datum A.

④ Complete the feature control callout; identify secondary datum callout;
② tolerance gage points appropriately (±0.03in one direction). Draw the

tolerance zone and explain design logic.

GENERAL TEST 4

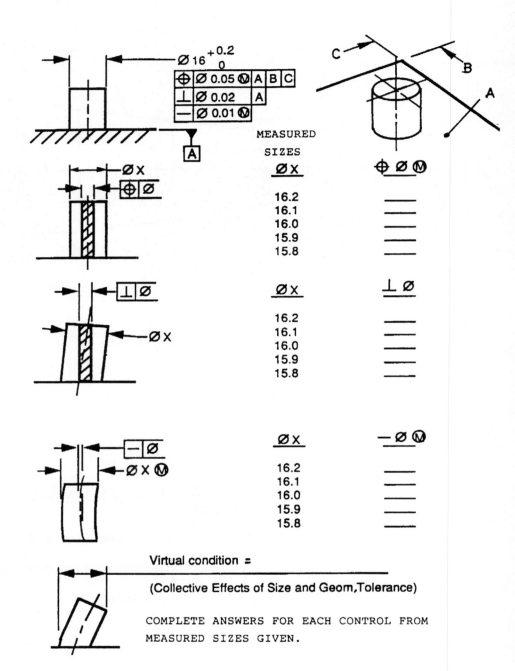

MEASURED
SIZES

ØX	⊕ Ø Ⓜ
16.2	—
16.1	—
16.0	—
15.9	—
15.8	—

ØX	⊥ Ø
16.2	—
16.1	—
16.0	—
15.9	—
15.8	—

ØX	— Ø Ⓜ
16.2	—
16.1	—
16.0	—
15.9	—
15.8	—

Virtual condition =

(Collective Effects of Size and Geom,Tolerance)

COMPLETE ANSWERS FOR EACH CONTROL FROM
MEASURED SIZES GIVEN.

GENERAL TEST 5

CHOOSE THE MOST APPROPRIATE ANSWER AT RIGHT. Not all available answers at the right are used, while some are used more than once. Question (7) has multiple answers.

_____ 1. Dimensions originating from a datum, control relationship of _____.

_____ 2. A feature is permitted ... the stated tolerance; no bonus as size varies...

_____ 3. A form control with a tolerance zone between two concentric circles.

_____ 4. This control is used to identify a tolerance of location.

_____ 5. This geometric control requires a basic angularity dimension.

_____ 6. LMC Symbol

_____ 7. Implied to apply RFS.

_____ 8. Condition of a part feature when it contains maximum amount of Material.

_____ 9. A tolerance zone between two parallel planes and parallel to a datum.

_____ 10. A tolerance zone between two parallel planes and perpendicular to a datum.

_____ 11. A geometric form control that has deviation from a straight line.

_____ 12. A tolerance zone between two concentric cylinders.

_____ 13. All datum planes intersect at right angles are 90°.

_____ 14. Datum identification symbol.

A. $\angle$

B. **Datum Ref. Frame**

C. **Form**

D. ——

E. (symbol)

F. $\boxed{A}$

G. .175

H. $\oplus$

I. Ⓜ

J. ◯

K. ◎

L. **Related Features**

M. **RFS**

N. //

O. Ⓛ

P. **Datum Target**

Q. $\perp$

GENERAL TEST 6

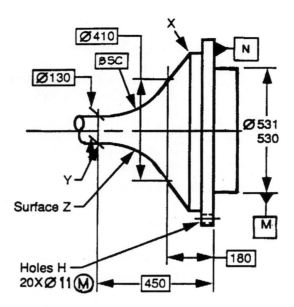

Illustrate:

1. Datum surface N to have a total runout of 0.2 FIM relative to datum ∅ M.
2. Surface Z:
    circular runout to be 0.5 relative to primary datum M and secondary datum N.

    line elements of surface Z to have a 0.3 tolerance from the true basic profile between points X and Y.
3. Hole H are to be in true position at MMC relative to secondary datum M at MMC and primary datum N. The fasteners are ∅M10 screws and the mating part has been assigned 60% of the available tolerance.

GENERAL TEST 7

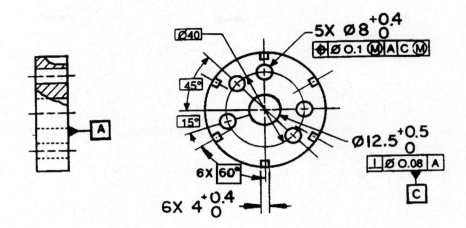

- Illustrate the 6 grooves to be located from primary datum A and secondary datum C (RFS) within 0.2 (MMC), and as a group, are datum B.
- Illustrate the 5 small holes also related to tertiary datum B (MMC).
- What is the MMC size of the small holes _____?
- What is the virtual condition of the small holes _____?
- Calculate a new positional tolerance of the small holes for a fixed fastener condition if the bolts are M6 _____.
- What is the virtual condition of the large center hole _____?
- What is the total *bonus* tolerance allowed for the 5 small holes_____?

GENERAL TEST 8

(Please indicate True or False beside each statement.)

1. Straightness per unit of length may be used if a feature must be controlled incrementally as well as over its entire length.

2. Flatness also controls straightness of surface line elements.

3. The circularity tolerance zone description is "a diameter."

4. A parallelism tolerance zone must be contained within the size tolerance of a feature surface.

5. A cylindricity tolerance zone may be defined as two concentric circles.

6. To control parallelism of surface line elements to a datum, the words **EACH ELEMENT** must be added beneath the feature control frame.

7. A concentric tolerance zone is described as "a radius."

8. Concentricity is used for the axial control of a feature of size to a datum axis.

9. A profile zone may be unilateral or bilateral.

10. The use of an axial straightness control may allow a cylindrical feature to violate the MMC perfect form boundary.

11. With the exception of position, geometric controls and related references apply RFS.

12. Runout or position controls are generally preferred over concentricity.

13. A coaxial relationship may be controlled by specifying a positional callout RFS with the datum reference also RFS.

14. LMC is most commonly used to control wall thickness.

15. True position is the theoretically exact location of a feature, axis or center plane.

16. In a feature control frame, datum references must be in order of: A primary, B secondary, and C tertiary.

17. Datum features are theoretically exact.

18. Cylindricity is applied MMC because the tolerance zone is complex.

GENERAL TEST 9

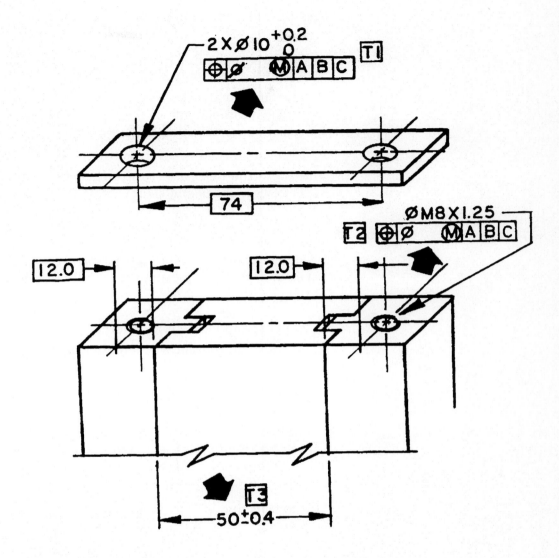

What is the tolerance for T_1 and T_2 if T_3 is a plus/minus tolerance?

GENERAL TEST 10

Functional Design Analysis

1. Determine the functional importance of a part (and features) within an assembly or sub-assembly.
2. Identify the attachment and operational transference features for each component.
3. Develop the priority of interface between features.
4. Determine the datum frameworks and feature controls.
5 Review the design for unfinished or incomplete controls, and for review of fits and clearances.

Attachment Features—Those features which contact and locate components within a given assembly.

Operational/Transference Features—Those features which perform or transfer a function, usually in relation to other aspects of the design:

a. support and location for another part
b. operational contact with another part

Interfacing features have a priority of contact based on fits and clearances, in addition to function. This priority is the principle behind Primary, Secondary, and Tertiary Datum frameworks. Primary contact gives the greater degree of control, with any feature error (squareness, etc.) transferred to the Secondary contact. The Tertiary contact "locks in" the framework.

Datum Selection

Datum selection depends on the size, form, orientation, profile, runout and location controls required for the correlation of mounting and function in a design.

Size and form describe the individual features within a component which become datums.

Orientation, runout, and location describe feature datum relationships, once datums are determined.

Profile can be used with or without such a relationship.

Applying Feature Controls

1. Determine the operational/transference features in the design. These features will also require form, orientation, runout, and location controls.

Determine primary attachment feature with necessary form controls.

2. Determine secondary feature with necessary orientation or runout controls (attitude error will occur in the secondary datum).

3. Determine tertiary attachment feature, if present. The tertiary feature will also have attitude error relative to the primary datum. Determine necessary position or orientation controls.

   Review completeness: fits, clearances, tolerance, and feature control impact on mating features and assembly.

   Complete, as far as possible the following design requirements: datums, feature controls, and tolerances.

GENERAL TEST 10A

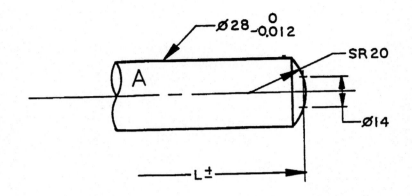

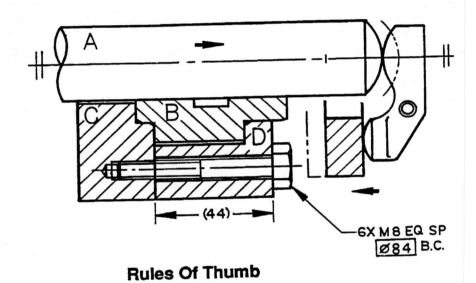

Rules Of Thumb

- Form Tolerance - (1/2 Size Tolerance) MAX

GENERAL TEST 10B

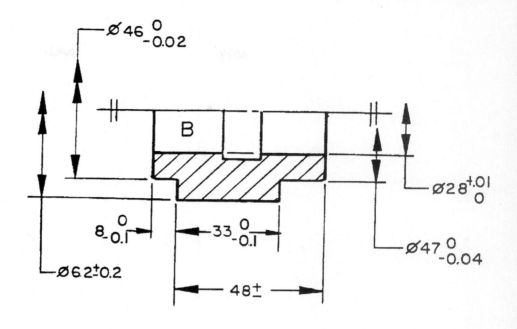

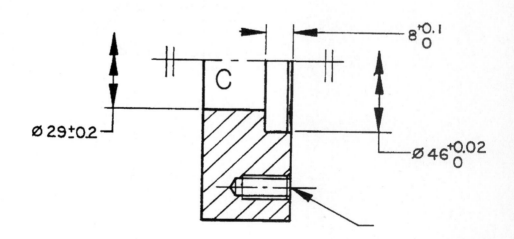

GENERAL TEST 10C

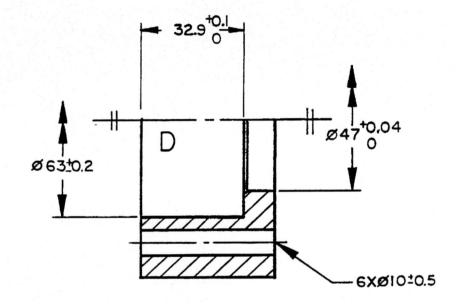

Glossary

Actual Size—measured size, includes local and mating size.

Actual Local Size—measured size at any cross section of a feature.

Actual Mating Size—value of minimum/maximum mating envelope.

Angularity—surface, axis, or centerplane at any angle (other than 90°) relative to a datum.

ANSI—American National Standard Institute.

ASME—American Society of Mechanical Engineers.

Basic—a numerical value that describes an exact theoretical size, shape, location, etc. of a feature, datum, or target. The value is placed within a box or rectangle.

Basic Size—that size from which limits and tolerances are derived.

Bilateral Tolerance—a tolerance which exists in two directions from a specified dimension.

Bonus Tolerance—a tolerance that may be added due to size variation.

Boundary, Inner—smallest possible feature envelope (cylinder) for an internal feature at MMC or external feature at LMC.

Boundary, Outer—largest possible feature envelope (cylinder) for an external feature at MMC or internal feature at LMC.

Calipers—tool to make opposed measurements.

Centerline—middle or mediam line of feature.

Centerplane—middle or median plane of feature.

Chamfer—edge break or cut of a feature shoulder to remove sharp edges.

Circular Runout—composite surface control at the cross section of a feature relative to a datum axis.

Circularity—form control for the surface elements (cross section) of a feature of revolution (cylinder, cone, sphere). Also called roundness.

Clearance Fit—condition with the internal part size limit is always smaller than the external mating part size limit.

Coordinate Measuring Machine (CMM)—device that measures, records, calculates, and analyzes measured feature variations.

Coaxiality—condition of two or more features sharing the same centerline.

Composite Tolerance—tolerance providing a relationship to a datum framework, as well as feature relationship (i.e., profile/position).

Concentricity—condition where one or more features of various possible shapes are in line with a datum feature axis, both/all RFS.

Controlled Radius—radius without flats or reversals within size limits.

Contour—*see* Profile.

Coplanarity—two or more surfaces that lie in the same plane.

Counterbore—stepped increase in feature hole diameter.

Cylindricity—form control with all surface elements equadistant from a common axis.

Datum—theoretically exact point, line (axis), or plane.

Datum Axis—axis established by the datum feature.

Datum Feature—actual physical feature (surface) on a part.

Datum Feature of Size—physical feature that has size variation.

Datum Feature Symbol—a reference letter contained within a box.

Datum Plane—theoretically exact plane.

Datum Reference Frame(work)—three mutually perpendicular planes.

Datum Simulator—processing or inspection equipment feature surfaces.

Datum Surface—actual feature surface (flat surface, hole, slot, etc.) used to establish the theoretically perfect datum.

Datum Target—points, lines, or areas used for consistency and repeatability of datum features.

Datum Target Symbol—large circle containing letters and numbers for identification of targets.

Diameter Symbol—circle with a diagonal line.

Dimension—numerical value used with lines, notes, symbols, etc. to define part characteristics.

Dial Indicator—device to measure variation from any desired condition.

Envelope, Actual Mating:

> *External*—smallest size that can be circumscribed about the feature so that it contacts the high points of the feature surface.

> *Internal*—largest size that can be inscribed within the feature so that it contacts the high points of the feature surface.

Extension Line—line to extend dimensional features.

Feature—general term for a physical portion of a part (hole, surface, slot, thread, etc.).

Feature Axis—the axis of the feature's true geometric counterpart.

Feature Centerplane—The centerplane of the features true geometric counterpart.

Feature Control Frame—rectangular boxes which form specification controls in symbol sentences.

Feature of Size—feature, that when changed, will effect the physical weight of a part such as holes, length, width, or thickness.

Fixed Fastener Condition—assembly condition of two or more parts with one part having features that are threaded, press fit or line fit in nature, with a fastener involved.

Fit—general term to identify looseness or tightness of assembly conditions.

Flatness—form control for a surface with all surface elements in one plane.

Floating Fastener Condition—assembly condition of two or more parts with assembly clearance in both parts.

Force Fit—*see* Interference Fit.

Form Tolerance—category of tolerances for control of individual features such as flatness, circularity, straightness, or cylindricity.

Free State Variation—condition of a part when unrestrained the dimensional limits may change.

Full Indicator Movement—difference between minimum and maximum limits read on an indicator device.

Functional Gage—a receiver gage that will receive the part with no force applied.

Geometric Characteristics—basic elements for control in geometric tolerancing such as form, orientation, profile, runout, and location controls.

Geometric Tolerance—application of geometric characteristic controls.

Implied Datum—unspecified datum implied by the dimensioning structure on the part drawing.

Interference Fit—condition where the size limits of two parts will always result in an interference or press fit.

Irregular Curve—curve that does not have a constant radius.

Key-way/Key-slot—slot to control fixed (rotational) relationships between parts.

Least Material Condition—feature of size contains the least material or when part weighs the least.

Limit Dimension—minimum and maximum dimensional limits are specified.

Limits of Size—specified maximum and minimum size.

Line Fit—condition with size limits which result in a zero tolerance.

Location Tolerance—specification allowing departure from perfect or exact location.

Material Condition—MMC, RFS, or LMC.

Maximum Dimension—the extreme acceptable upper limit dimension.

Maximum Material Condition (MMC)—condition where feature of size contains the most material or when the part weighs the most.

Median Line—line derived from the centerpoints of cross sections of a feature, say in the middle.

Median Plane—plane derived from the centerpoint lines of cross sections of a feature, say in the middle.

Micrometer—measuring device for opposed (diametral) measurements.

Minimum Dimension—the extreme acceptable lower limit dimension.

Modifier—term and/or symbol to describe appropriate material condition applications (RFS, MMC, or LMC).

Multiple Datum Reference Frames—condition where features are controlled to more than one set of reference datums.

Nominal Size—term for general identification.

Orientation Tolerance—category of tolerances that control one feature relationship (attitude) to another (angularity, parallelism, and perpendicularity).

Origin of Measurement—points, lines, or planes from which measurements originate or are taken.

Paper Gaging—graphical and mathmatical analysis of inspection data.

Parallelism—condition of feature axis, line element or surface relative to a datum axis or surface.

Perfect Form at MMC—extreme perfect form envelope, all dimensions at MMC.

Perpendicularity—condition of feature axis, line element, or surface at right angle to a datum axis or surface.

Position Tolerance—tolerance zone within which a feature axis or centerplane is allowed to vary.

Primary Datum—datum with primary influence in/on a set of design criteria. The first datum listed in the feature control frame.

Profile Tolerance—condition that allows a line or surface to vary by a uniform amount, either bilaterally or unilaterally, from the desired true profile. May control size, form, location, and orientation depending on its use.

Profile, All Around—method of applying a constant perimeter tolerance.

Projected Tolerance Zone—tolerance zone that is projected above/below feature surface, normally equal to thickness of mating part.

Reference Dimension—dimension without tolerance, enclosed by paren ( ), for information only, not intended for application by inspection or manufacturing.

Regardless of Feature Size (RFS)—condition that indicates a feature tolerance applies regardless of the feature size, within the size limits.

Resultant Condition—a variable boundary generated by collective effects of size, size bonus, tolerance, geometric tolerance, and bonus tolerance which occurs as a feature departs from the specified material condition. Generally considered opposite of virtual condition.

Roundness—*see* Circularity.

Rules—preconditions, givens.

Roughness—*see* Surface Texture.

Runout—*see* Circular and Total Runout.

Runout Tolerance—composite or total value measured at a feature surface and measured relative to a datum axis through one full revolution of the part on the datum axis.

Secondary Datum—datum with lesser influence on function or design criteria. The second datum listed in the control frame.

Simulated Datum—datum with lesser influence on function or design criteria. The second datum listed in the control frame.

Simultaneous Datum Features—two features, when used together make a single datum plane or axis.

Size Dimension—value that defines a feature of size.

Size Feature—*see* Feature of Size.

Size Tolerance—states allowed departure from desired size.

Spherical Features—feature with all points on the surface equidistant from a single point.

Squareness—*see* Perpendicularity.

Statistical Tolerance—mathmatical manipulation of data.

Straightness—form control where line elements of a surface or a feature axis is a straight line.

Surface Texture—primary (roughness) and secondary (waviness) surface tool marks generated by processing. Also includes lay of tool marks.

Symmetry—condition of a feature equally disposed about the centerplane of a datum feature.

Tangent Plane—plane that contact the high points of a surface.

Taper—diametral change per unit length of a feature.

Target—*see* Datum Target.

Tertiary Datum—datum with least influence on design criteria. The third datum reference in the control frame.

(Times) Places—number of occurrences of a feature or dimension, indicated by an "X."

Tolerance—total permissible variation of a specification.

Total Runout—composite control of all surface elements of an entire feature surface relative to a datum axis.

Transition Fit—condition where the size limits of two parts may result in either a clearance or interference.

True Geometric Counterpart—theoretically perfect feature, or datum feature, virtual condition (or actual mating envelope) boundary.

True Position—theoretically exact location in relation to a datum reference or other feature.

Unilateral Tolerance—tolerances that exists in one direction from a specified dimension.

Virtual Condition—condition created by combined effects of size (MMC or LMC) and any geometric tolerance.

Waviness—*see* Surface Texture.

Workpiece—part or assembly in process or evaluation.

Zero Tolerancing—tolerancing technique in which tolerance is allowed based on a feature departure from MMC or LMC size limits only.

Answers to Exercises

EXERCISE 1-1. GENERAL DIMENSIONING

1. General Rule #1 controls:
    a. form as well as size of feature.   Ⓣ   F
    b. the relationships of shaft ends to the diameter.   T   Ⓕ
2. General Rule 2 implies RFS is applied to runout controls.
3. Prior to ANSI Y14.5–1973, position controls implied MMC.   Ⓣ   F
4. ISO drawings imply MMC for position controls.   T   Ⓕ
5. All dimensions on a drawing must have a tolerance.   T   Ⓕ
    (ref. min, max)
6. Drawing callouts referring to screw threads apply to the pitch
    diameter.   Ⓣ   F
7. Identify the figures in the feature controls frame below:

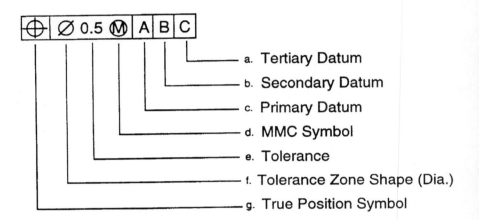

a. Tertiary Datum
b. Secondary Datum
c. Primary Datum
d. MMC Symbol
e. Tolerance
f. Tolerance Zone Shape (Dia.)
g. True Position Symbol

8. Identify each of the geometric characteristic symbols below.

— Straightness
// Parallelism
▱ Flatness
⊥ Perpendicularity
○ Roundness
∠ Angularity
◎ Concentricity

Cylindricity
Circular Runout
Profile Line
Position
Total Runout
Profile Surface
Symmetry

EXERCISE 2-1. CIRCULARITY (ROUNDNESS)

1. Circularity is applied MMC unless otherwise specified.   T   (F)
2. Circularity tolerance controls apply at the <u>surface or cross section</u> of a cylindrical shape.
3. Circularity controls *may* be applied to a tapered shaft.   (T)   F
4. Circularity is always datum related.   T   (F)
5. Circularity is a refinement of size, therefore, a form control.   (T)   F
6. Vee blocks are best for circularity measurement.   T   (F)
7. Circularity tolerance must be within size limits.   (T)   F
8. The use of MMC will allow a bonus circularity tolerance.   T   (F)
9. Indicate a circularity callout with a 0.1 tolerance in the figure below. Draw the tolerance zone.

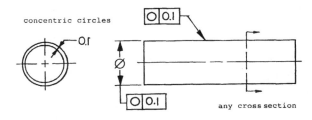

EXERCISE 2-2. CYLINDRICITY

1. Cylindricity is applied MMC unless otherwise specified.   T   (F)
2. Cylindricity is applied to cylindrical shapes and must be larger than size tolerances.   T   (F)
3. Cylindricity controls ○ circularity and — <u>straightness</u> of surface elements.
4. Cylindricity is always datum related.   T   (F)
5. Cylindricity tolerance zone is the space between two concentric circles.   T   (F)
6. A cylindricity tolerance may extend beyond the MMC size envelope.   T   (F)
7. Cylindricity may be used to control taper in the figure below.   (T)   F

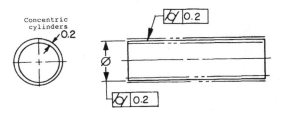

8. Indicate a cylindricity callout with a 0.2 tolerance in the figure above. Draw the tolerance zone.

EXERCISE 2-3. FLATNESS

1. Flatness is applied MMC.   T   (F) (Can't modify a surface)
2. Flatness also controls ___ of surface elements.
3. Flatness tolerance is additive to size tolerances.   T   (F)
4. Flatness is always datum related.   T   (F)
5. If flatness controls an entire surface, then it also controls the squareness of the surface to a surface plate.   T   (F)
6. The figure below *may* have a flatness tolerance of 0.5.   (T)   F (*Note:* PERFECT FORM AT MMC NOT REQUIRED.)

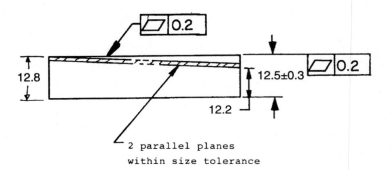

7. Can MMC be applied to the *surface* in the above figure?   Yes   (No)
8. Indicate a flatness callout (on the top surface) with a 0.2 tolerance in the figure above. Draw the tolerance zone.

EXERCISE 2-4. STRAIGHTNESS

1. Straightness is applied RFS unless otherwise specified. Ⓣ
   F
2. Straightness may be applied to surface <u>elements</u>, cylindrical <u>surface elements</u>, or <u>axes</u>.
3. Virtual condition is invoked when straightness is applied to a feature <u>axis</u>.
4. Straightness is always datum related.   T   Ⓕ

Complete the table below for the measured sizes given. Draw the tolerance zone.

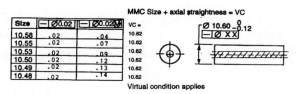

Size	⊟|Ø0.02|	⊟|Ø0.02Ⓜ|	VC =
10.58	.02	.04	10.62
10.55	.02	.07	10.62
10.53	.02	.09	10.62
10.50	.02	.12	10.62
10.49	.02	.13	10.62
10.48	.02	.14	10.62

MMC Size + axial straightness = VC

Ø 10.60 −0.12/−0

Virtual condition applies

Complete the table below for the measured sizes given. Draw the tolerance zone.

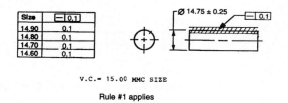

Size	⊟|0.1|
14.90	0.1
14.80	0.1
14.70	0.1
14.60	0.1

Ø 14.75 ± 0.25   ⊟\|0.1\|

V.C.= 15.00 MMC SIZE

Rule #1 applies

EXERCISE 2-5. TRUE/FALSE

With Straightness of Surface Elements:

1. Ⓜ or Ⓛ may not be applied.   T
2. Perfect form at MMC is required.   T
3. Straightness tolerance must lie within size tolerance.   T
4. Virtual condition will allow Rule #1 to be violated.   F
5. RFS is understood to apply.   T

With Straightness of an Axis:

1. Datums are applied.   F
2. Rule #1 does not apply when straightness is applied to a feature axis or centerplane.   T
3. Virtual condition is equal to MMC size plus axial straightness tolerance.   T
4. With axial straightness, the diameter symbol is optional.   F
5. The mating size (virtual condition) of a shaft that is 15mm diameter ±0.3mm, with an axial straightness of 0.3 diameter is 15.3mm.   F

EXERCISE 3-1. DATUMS

1. In order of importance datums are: <u>primary</u>, <u>secondary</u>, or <u>tertiary</u>.

2. Occasionally, datum features are not possible or practical to use. In this case, <u>datum targets, simulated datums, or partial</u> datums are defined.

3. Ideally, datums should be selected and specified as determined by <u>their functional importance</u>.

4. The figure below represents a <u>co-planar</u> datum.

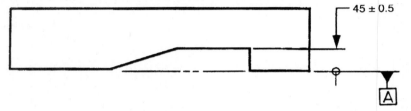

5. The datum which "stops" rotation of a cylindrical part is normally the <u>tertiary</u> datum.

6. The use of opposite ends of a shaft simultaneously for datum purposes is shown by this callout:

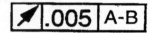

7. The datum framework consists of _3_ mutually <u>perpendicular</u> planes.

8. In a functional receiver gage (MMC), secondary and tertiary datums that are features of size are applied at their <u>virtual condition</u>.

In the figure to the right:

- Specify the mounting surface in the right view as datum A.

- Specify the flat surface at the bottom as datum B.

- Specify the ⌀ 20 as datum C

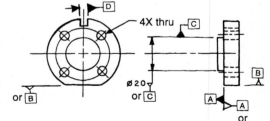

- Specify the slot centerplane as datum D.
- Which datum features above are subject to size variation and virtual condition? <u>C, D</u>
- Of datums A, C, and D, what sequence and modifier symbols appear the most logical, if functional gaging is to be used? Why?

⊥ Loc. Orient.

Why? A gives perpendicularity control; C centers the pattern of holes; and B orients the pattern of holes

True–False

1. Datum accuracy and accessibility are equally important.   T
2. Datum order is determined by functional influence.   T
3. The datum reference frame provides stability for measurement.   T
4. Datum planes have length, width, and depth.   F
5. Actual part surfaces are called datum features.   T
6. Datum symbols should be indicated to centerlines and centerplanes.   F

In your judgment (using the figure below):

- What is the primary datum? **(A)**
- What are the secondary and tertiary datums? **(B, C)**
- How would you complete the control callout for the oil hole and adjusting screw hole if functional gaging is to be used?
- Show the secondary datum to be perpendicular to the primary datum within 0.5 RFS.

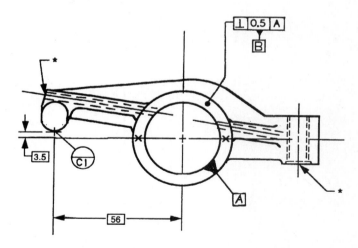

EXERCISE 4-1.   PARALLELISM

1. Parallelism is applied MMC unless otherwise specified.   T   ⒡
2. Parallelism is datum related.   ⓣ   F
3. On a flat surface, parallelism also controls ▱ (flatness).
4. Parallelism should be __less__ than size tolerance. (Use less than, in addition to, or greater than.)
5. A parallelism callout specifies a ± tolerance zone.   T   ⒡

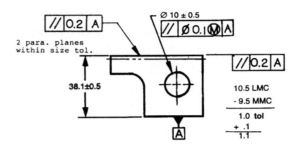

6. On the figure above, indicate a parallelism tolerance of 0.2 to datum A. Draw the tolerance zone (top surface).
7. Indicate the axis of the hole to be parallel to datum A within 0.1 diameter at MMC.
8. What is the size of the hole tolerance zone in the figure above if the diameter is at LMC? ∅ 1.1mm
9. What is the maximum (total) value the 38.1 dimension can be? __38.6__
10. What is the hole virtual condition *size* limit at MMC? ∅ 9.4
11. When applied to a flat surface, the parallelism tolerance is in addition to size tolerance.   T   ⒡

EXERCISE 4-2.   PERPENDICULARITY

1. Perpendicularity is applied RFS unless otherwise specified. (T)   F

2. Perpendicularity is datum related.   (T)   F

3. Virtual condition is invoked when perpendicularity is applied to a feature <u>axis</u> or <u>centerplane</u>.

4. Perpendicularity applied to a flat surface also controls ⟋ (flatness) as well as ― of surface elements.

5. Perpendicularity may be applied to cylindrical feature relationships.   (T)   F

6. The figure below requires a 90° angle dimension.   T   (F)

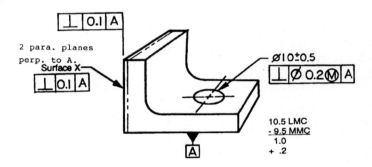

7. On the figure above, indicate surface X to be perpendicular to datum A within 0.1. Draw the tolerance zone.

8. Indicate the hole to be perpendicular to datum A within a diameter of 0.2 at MMC.

9. What is the size of the tolerance zone in the figure above if the feature is 10.5 diameter? <u>⌀ 1.2</u>

10. What is the virtual condition size of the hole at MMC? <u>⌀ 9.3</u>

EXERCISE 4-3.   ANGULARITY

1. Angularity is applied MMC unless otherwise specified.   T   (F)
2. Angularity is not datum related.   T   (F)
3. An angularity tolerance zone is: (a) (2) parallel planes or (b) a wedge shaped zone.
4. Angularity may be applied to a surface at MMC.   T   (F) (Can't modify a surface)
5. Angularity should be used to control the relationship of two 90° surfaces.   T   (F)
6. To control the hole in the figure below, a feature of size, the use of MMC will allow a bonus tolerance.   (T)   F
7. For the figure below, indicate the hole to have a 60° relationship to datum A, and provide an angularity callout of 0.1 diameter to datum A. Draw the tolerance zone.

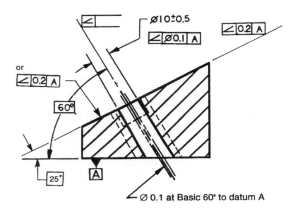

Ø 0.1 at Basic 60° to datum A

8. The LMC hole size is ∅ 10.5.
9. What is the virtual condition size of the hole at MMC? 9.4
10. Illustrate that the inclined surface is 25° basic relative to datum surface A, with an angularity tolerance of 0.2 to A.
11. Illustrate the tolerance zones in the figures below.

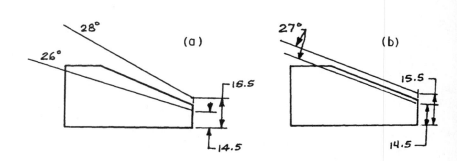

EXERCISE 5-1.   PROFILE (LINE/SURFACE)

1. Profile is applied RFS unless otherwise specified.   Ⓣ   F
2. Profile tolerancing is always datum related.   T   Ⓕ
3. Profile also may be used to control the ◯, ⬭, or taper of cylindrical surfaces or co-planar relation of flat surfaces.
4. ⌒ is the symbol for surface (depth) control.   T   Ⓕ
5. Profile may be used in combination with other geometric controls or with size controls.   Ⓣ   F
6. In certain applications, the profile tolerance zone may extend beyond size limits.   T   Ⓕ
7. A datum or datums may be used to orient the tolerance zone to a surface or surfaces.   Ⓣ   F
8. The profile tolerance zone is understood to be unilateral.   T   Ⓕ
9. Indicate the surface (between points X and Y) in the figure below to be relative to datum A primary and datum B secondary within 0.2. Draw the tolerance zone.

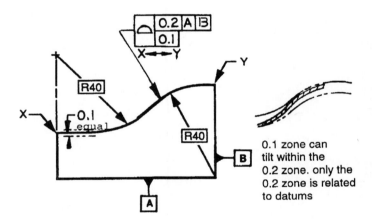

0.1 zone can
tilt within the
0.2 zone. only the
0.2 zone is related
to datums

10. Indicate the true profile of the surface in the figure above to be within 0.1 between X and Y.

EXERCISE 5-2.   SPECIFYING PROFILE OF A SURFACE BETWEEN POINTS

1.   Complete the drawing control callouts ① and ② to satisfy the requirements indicated.

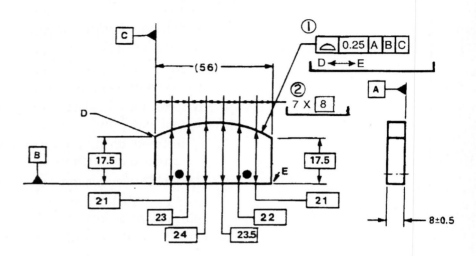

The surface between points D and E must lie between two profile boundaries 0.25 apart, perpendicular to datum plane A, equally disposed about the true profile and positioned with respect to datum planes B and C.

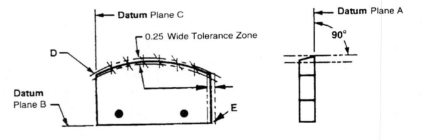

EXERCISE 6-1.   RUNOUT (CIRCULAR/TOTAL)

1. Runout is understood to apply at MMC.   T   (F)
2. Circular and total runout callouts control <u>surface</u> elements and include the errors of <u>form</u> and <u>location</u>.
3. Additionally, total runout controls O, ⌀ —, <u>and taper</u> of a cyclindrical surface.
4. Runout may be used with or without datums.   T   (F)
5. Runout controls are read at a feature surface and are relative to a datum axis.   (T)   F
6. The control of a convex or concave *surface* to a datum axis can be accomplished with circular runout.   T   (F)

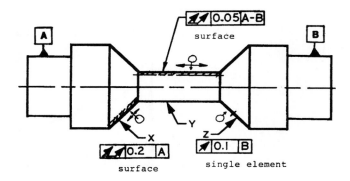

For the figure above:

7. Indicate a total runout to datum A of 0.2 for surface X.
8. Indicate surface Z to have a circular runout to datum B of 0.1.
9. Indicate surface Y to have a total runout to datum A and B of 0.05.
10. Draw the tolerance zone(s).
11. Runout may be applied to a surface, relative to a primary flat datum surface and secondary datum axis.   (T)   F

EXERCISE 7-1.   POSITION TOLERANCE

1. True position controls are implied MMC.   T   Ⓕ
2. A true position tolerance zone is a ± zone.   T   Ⓕ
3. Ⓜ allows a bonus tolerance.   Ⓣ   F
4. True position controls require use of datums for hole *pattern* controls.   Ⓣ   F
5. Indicate the correct position tolerance associated with each of the following hole sizes for the object shown below:

Size	Tolerance
8.4	0.2
8.5	0.3
8.6	0.4

6. The virtual condition for this hole is __8.2__ .

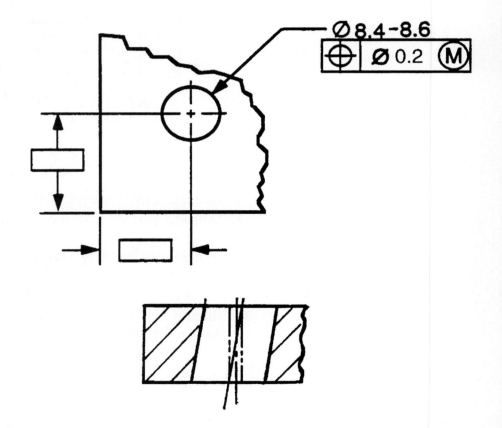

EXERCISE 7-2.   FORMULAS

What are the formulas for:

1. Floating fastener position tolerance.

$$T = H - F$$

2. Fixed fastener position tolerance.

$$T = \frac{H - F}{2}$$

3. Hole diameter virtual condition at MMC.

VC = size at Ⓜ − Geometric Tolerance

4. Shaft diameter virtual condition at MMC.

VC = size at Ⓜ + Geometric Tolerance

EXERCISE 7-3.   APPLYING THE FORMULAS

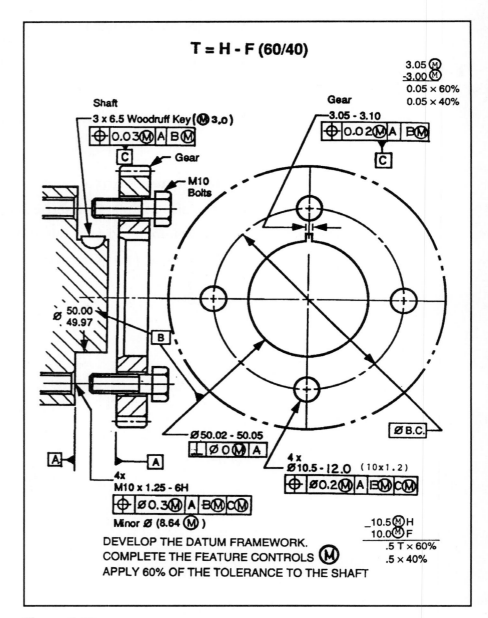

Figure 7-48.

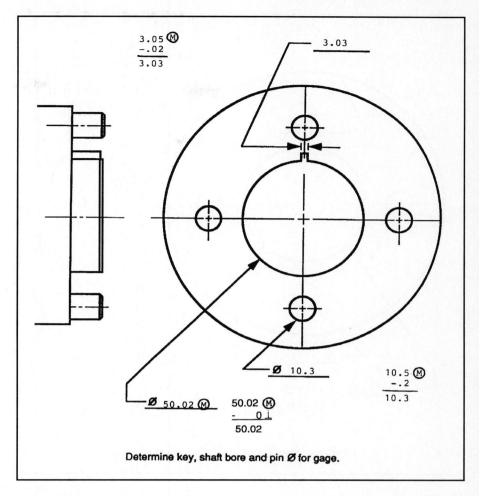

$$\frac{3.05\ \text{Ⓜ}}{3.03}$$
−.02

3.03

$$\frac{10.5\ \text{Ⓜ}}{10.3}$$
−.2

Ø 10.3

Ø 50.02 Ⓜ   50.02 Ⓜ

$$\frac{- \quad 0\ \perp}{50.02}$$

Determine key, shaft bore and pin Ø for gage.

Figure 7-49.   Gear gage.

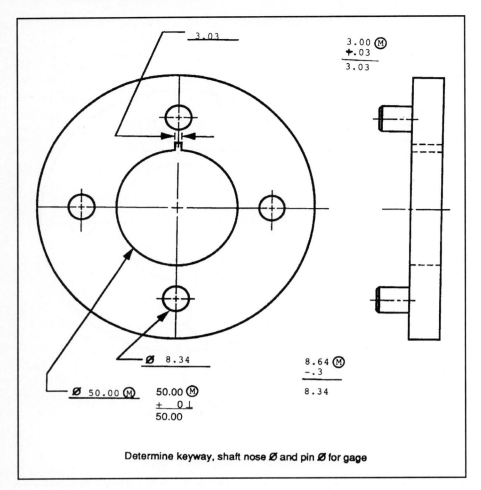

Determine keyway, shaft nose Ø and pin Ø for gage

Figure 7-50.   Shaft gage.

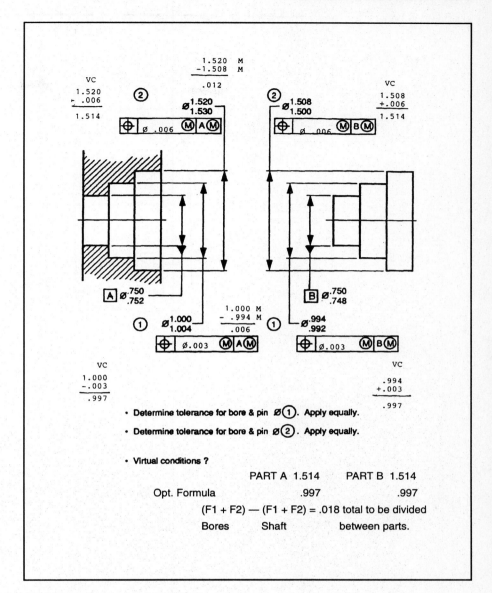

- Determine tolerance for bore & pin Ø①. Apply equally.
- Determine tolerance for bore & pin Ø②. Apply equally.

- Virtual conditions ?

	PART A 1.514	PART B 1.514
Opt. Formula	.997	.997

(F1 + F2) — (F1 + F2) = .018 total to be divided
Bores   Shaft   between parts.

Figure 7-51.

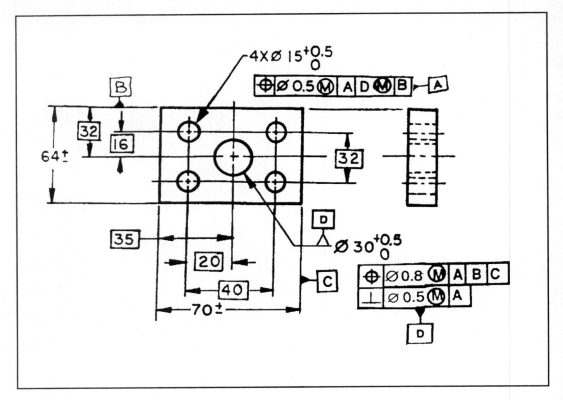

Figure 7-53.

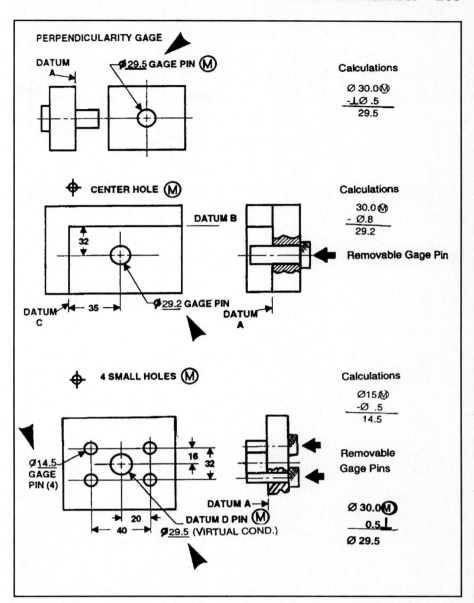

Figure 7-54.

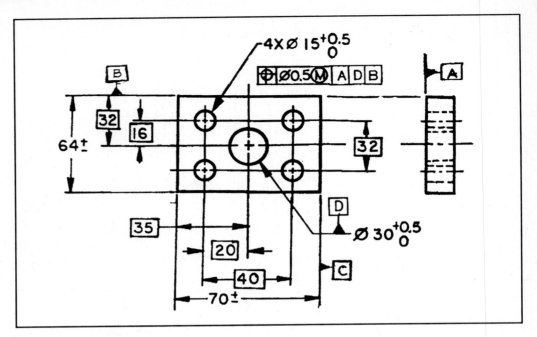

- Adjust the callout so the 4 small holes are MMC, but relative to the large hole RFS, and tertiary datum B.

- Tolerance zones are similar to the previous exercise except as noted.

Figure 7-55.

EXERCISE 7-4. CONCENTRICITY

1. Concentricity is always datum related. (T)   F
2. Concentricity is applied <u>RFS</u>.
3. The concentricity tolerance exists at the feature <u>axis (median line)</u> and relates to the datum <u>axis</u>.
4. Therefore, the use of V-blocks for datum set-up is recommended. T (F)
5. Concentricity controls require the verification of feature axes, without regard to surface conditions, to datum axes. Therefore, unless there is an exact need for this control, the use of position or runout is recommended. (T)   F
6. Indicate the middle and smaller diameters of the object in the figure below to be concentric with the larger diameter within 0.1. Draw the tolerance zone.

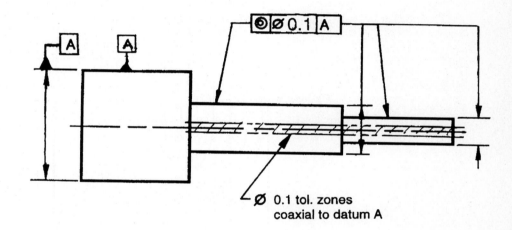

Ø 0.1 tol. zones
coaxial to datum A

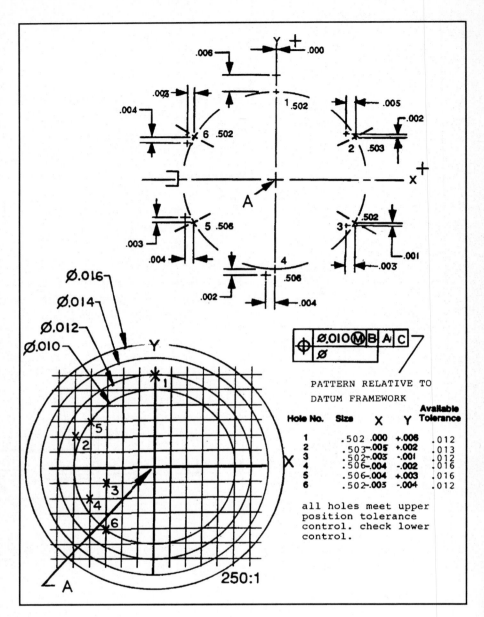

Figure 8-8.

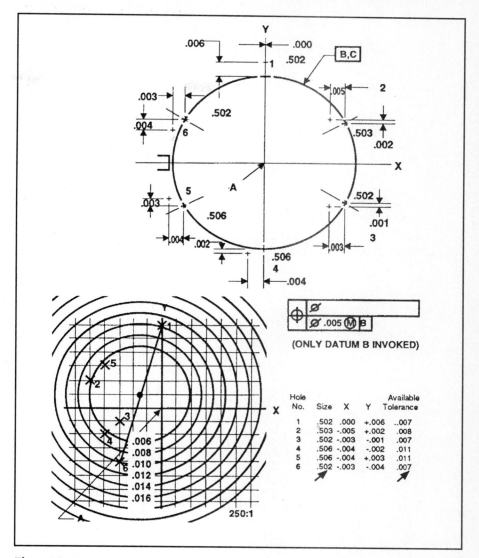

Figure 8-9.

Index